Significant Changes to the NEC

2005 EDITION

Significant Changes to the NEC

2005 EDITION

NJATC

THOMSON

DELMAR LEARNING™

Australia   Canada   Mexico   Singapore   Spain   United Kingdom   United States

DELMAR

THOMSON LEARNING

Significant Changes to the NEC, 2005 Edition

NJATC

Vice President, Technology and Trades SBU:
Alar Elken

Executive Director, Professional Business Unit:
Gregory L. Clayton

Product Development Manager:
Patrick Kane

Executive Marketing Manager:
Beth A. Lutz

Channel Manager:
Erin Coffin

Marketing Coordinator:
Penelope Crosby

Production Director:
Mary Ellen Black

Production Manager:
Larry Main

Senior Project Editor:
Christopher Chien

Art/Design Coordinator:
Francis Hogan

Editorial Assistant:
Sarah Boone

NOTICE TO THE READER

Contents

CHAPTER 4
Equipment for General Use, Articles 400–490   **204**

CHAPTER 5
**Special Occupancies,
Articles 500–590**   **264**

**CHAPTER 6
Special Equipment,
Articles 600–695**   **318**

Preface

The National Electrical Code (NEC®) is the most widely recognized and accepted electrical standard in the world. Every three years, the NEC is revised to reflect the newest installation practices utilized by the electrical industry. In addition, during the past several Code cycles, a task group has worked to systematically increase the usability of the Code and restructure the document for both new students of the Code and those who have long relied upon it in the field.

The purpose of *Significant Changes to the NEC—2005 NJATC Version* is to familiarize electricians, electrical contractors, electrical inspectors, and electrical engineers with the most important changes in the 2005 NEC. From the designing engineer to the installing electrician, everyone in the industry plays a vital role in ensuring that installations conform to the NEC and are performed in a manner that provides the very highest level of safety for the customer and the public. This text is designed to assist those who have this responsibility in understanding the actual Code changes, and more importantly, the reason behind the changes.

For the most part, this text is arranged to follow the general layout of the NEC, including Code articles and section number format. When reviewing changes to the NEC, remember that all revisions are indicated by a vertical line that appears in the margin of the page.

Throughout this book, each change is accompanied by either a photograph or an illustration to assist and enhance the reader's understanding of the specific change. A summary and a discussion of the significance of the change are also provided.

As with any Code change text, *Significant Changes to the NEC—2005 NJATC Version* is best used as a study companion to the 2005 NEC. Readers should first read the "Change Summary." Next, they should read the actual Code reference as it appears in the "NEC 2005 Change" section. Finally, they should review the "Change Significance" section to understand the intent and impact of the revision.

Instructors who plan to use this text for a course should attempt to attend the National Joint Apprenticeship and Training Committee

(NJATC) *Significant Changes to the NEC—2005 NJATC Version* Train-The-Trainer (TTT) course. This 2-1/2 day course provides a venue to discuss the 2005 changes in depth with industry members who were directly involved in updating the 2005 NEC. In addition to the TTT course, a CD-ROM that contains all of the pictures and illustrations in the text is available to assist in the delivery of this course.

Undertaking a serious study of the 2005 NEC can be both a challenging and rewarding experience. This text has been designed by the NJATC to assist students in their personal pursuit of a solid understanding and application of the requirements contained in the 2005 NEC.

Acknowledgements

The NJATC and Thomson Delmar Learning would like to thank the following technical editors for their valuable contributions to the development of this book:

Mark E. Christian
James T. Dollard Jr.
Palmer Hickman
Donald M. King
Jamie McNamara
Harry Ohde
Phil Simmons
Janet D. Skipper

Significant Changes to the NEC

2005 EDITION

CHAPTER

1

Annex G, Articles 90, 100, and 110

Annex G

Administration and Enforcement
NEC Page 731

CHANGE TYPE: Relocation (Proposal 1-6)

CHANGE SUMMARY: Article 80 from the 2002 NEC on Administration and Enforcement is relocated to Annex G.

2005 CODE: Annex G

CHANGE SIGNIFICANCE: The substantiation for moving Article 80, "Administration and Enforcement," to Annex G states that the information in the article is not intended for mandatory enforcement and is not assigned to any Code-Making Panel (CMP). Also, the information in Article 80 should not be adopted as a part of the NEC unless the local jurisdiction does not have an electrical inspection jurisdiction in operation. It is more appropriately included as an annex.

In reality, the information in Article 80 constitutes a model ordinance for creating and operating an electrical inspection program for a city, county, or other governmental jurisdiction. In all likelihood, a governmental agency would use the information in the article (now an annex) as part of an enacting ordinance rather than adopting it as part of the electrical code intended to regulate the installation of electrical equipment and systems.

Moreover, Article 90, "Introduction to the Code," should be the first article, since it provides the purpose, scope, and code arrangement for the whole Code.

FIGURE 1.1

Annex G
Administration and Enforcement

Annex G is not a part of the requirements of this NFPA document and is included for informational purposes only. This Annex is informative unless specifically adopted by the local jurisdiction adopting The National Electrical Code.

80.1 Scope. The following functions are covered:

(1) The inspection of electrical installations as covered by 90.2
(2) The investigation of fires caused by electrical installa–

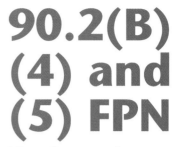
90.2(B) (4) and (5) FPN
Not Covered
NEC Page 24

CHANGE TYPE: New (Proposal 1-25; Comment 1-9)

CHANGE SUMMARY: A Fine Print Note (FPN) has been added regarding the typical organization of utilities and the application of regulatory bodies, the NEC, and other codes to the utilities.

2005 CODE: "**FPN to (4) and (5):** Examples of utilities may include those entities that are typically designated or recognized by governmental law or regulation by public service/utility commissions and that install, operate, and maintain electric supply (such as generation, transmission, or distribution systems) or communication systems (such as telephone, CATV, Internet, satellite, or data services). Utilities may be subject to compliance with codes and standards covering their regulated activities as adopted under governmental law or regulation. Additional information can be found through consultation with the appropriate governmental bodies, such as state regulatory commissions, the Federal Energy Regulatory Commission, and the Federal Communications Commission."

CHANGE SIGNIFICANCE: This FPN adds an explanation of local, state, or federal agencies that regulate typical utilities to help the user understand the Code. The FPN applies to both communications utilities under 90.4(B)(4) and to electrical utilities under 90.2(B)(5).

These sections provide details of which portions of the utilities' systems or plants are not covered by the Code. Section 90.2 does not intend to regulate entities entitled to make electrical installations or the types of licenses required but whether the installation of conductors, equipment, and systems are covered by the Code.

FIGURE 1.2

The FPN explains that utilities are typically designated or recognized by government law or public service utility commission regulations. These utilities install, operate, or maintain the relevant electric supply and communications systems. For electric utilities, systems may include generation, transmission, or distribution of electric energy. Communications utilities typically supply telephone, CATV, Internet, satellite, or data services.

CHANGE TYPE: New Paragraph (Proposal 1-34)

CHANGE SUMMARY: Brackets that contain section references to another National Fire Protection Association (NFPA) document are intended for informational purposes only and are not enforceable. They indicate the source of the extracted text and immediately follow the extracted text.

2005 CODE: "Brackets containing section references to another NFPA document are for informational purposes only and are provided as a guide to indicate the source of the extracted text. These bracketed references immediately follow the extracted text."

CHANGE SIGNIFICANCE: A new paragraph has been added to clarify the purpose of brackets that indicate the source of information extracted from other NFPA standards. (The 2002 NEC used brackets to identify the source of the extracted text; this new paragraph explains the purpose of those bracketed references.)

Information in brackets, such as "[NFPA 33, 1.6 Definitions; NFPA 34, 1.6 Definitions, 4.2.1, 4.2.2, 4.3.1]," identifies the parent document for the information that is extracted and placed in the NEC. The information, if changed, is modified to fit NEC style without changing the requirement in any way. Should the NEC user want to revise the requirements in the extracted text, a proposal must be made to the parent document.

90.5(C)
Explanatory Material
NEC Page 24

FIGURE 1.3

695.3 Power Source(s) for Electric Motor-Driven Fire Pumps. Electric motor-driven fire pumps shall have a reliable source of power.

(A) Individual Sources. Where reliable, and where capable of carrying indefinitely the sum of the locked-rotor current of the fire pump motor(s) and the pressure maintenance pump motor(s) and the full-load current of the associated fire pump accessory equipment when connected to this power supply, the power source for an electric motor-driven fire pump shall be one or more of the following.

(1) Electric Utility Service Connection. A fire pump shall be permitted to be supplied by a separate service, or from a connection located ahead of and not within the same cabinet, enclosure, or vertical switchboard section as the service disconnecting means. The connection shall be located and arranged so as to minimize the possibility of damage by fire from within the premises and from exposing hazards. A tap ahead of the service disconnecting means shall comply with 230.82(5). The service equipment shall comply with the labeling requirements in 230.2 and the location requirements in 230.72(B). [NFPA 20:9.2.2]

Article 100

Bonding Jumper, System

NEC Page 26

CHANGE TYPE: New (Proposal 1-63; Comment 1-38)

CHANGE SUMMARY: A new definition has been added to describe the conductor used in separately derived systems for grounding and bonding the system or equipment. (The Technical Correlating Committee chose to locate the term in Article 100 rather than in Article 250, even though the term is used only in Article 250.)

2005 CODE: "**Bonding Jumper, System.** The connection between the grounded circuit conductor and the equipment grounding conductor at a separately derived system."

CHANGE SIGNIFICANCE: This proposal adds a new definition similar to that of *Bonding Jumper, Main*, but the new term applies only to separately derived systems. The term is used in Article 250 to clearly identify the conductor that connects the derived system to ground and provides the fault current path.

By defining this term, CMP-5 is able to describe it clearly and can thus regulate the proper size and installation of the conductor.

FIGURE 1.4

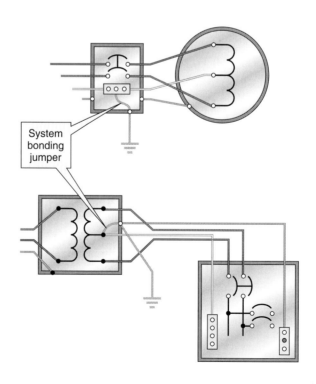

System bonding jumper

CHANGE TYPE: New (Proposal 1-76; Comment 1-55)

CHANGE SUMMARY: The Code adds a new definition to Article 100 for *coordination* (*selective*). (A similar term, *Coordination*, was previously used in 240.2 in the 2002 NEC.) The use of the term now suggests that only the overcurrent device closest to the fault will open, which will allow other circuits to operate without interruption.

2005 CODE: "**Coordination (Selective).** Localization of an overcurrent condition to restrict outages to the circuit or equipment affected, accomplished by the choice of overcurrent protective devices and their ratings or settings."

CHANGE SIGNIFICANCE: Selective coordination is accomplished by choosing overcurrent devices that do not have overlapping portions of their trip curve. Manufacturers of all overcurrent devices, be they circuit breakers or fuses, publish time-current curves that show the tripping characteristics of the overcurrent device. These time-current curves indicate the overload and short-circuit protection provided by the overcurrent device.

As illustrated in the example, a 400-ampere circuit breaker for a feeder supplies a panelboard. A 90-ampere breaker supplies a piece of equipment. The circuit breakers are operating in series. During the normal overload operation, the 90-ampere breaker will trip before and without the 400-ampere breaker opening. However, during a short circuit, both breakers will open. These breakers are not selectively coordinated. The reason for this is that during a short-circuit or ground-fault condition, enough current flows into the circuit to cause the short-circuit element of the circuit breaker to open.

Some manufacturers publish time/current curves on a grid system that can be overlaid using a light table to help select coordinated overcurrent devices. Other manufacturers publish coordination data to indicate "families" of fuses that will be selectively coordinated when used in a series. Software may also be available to determine selective coordination.

Selective coordination is required for hospitals and other buildings with critical care areas or that use electrical life support equipment in 517.17. Selective coordination is also required for elevators in 620.62 where a single feeder supplies more than one disconnecting means. A new requirement for selective coordination of emergency system overcurrent devices has been added as 700.27 and, for legally required standby systems, as part of 701.18. Selective coordination is not required for optional standby systems in Article 702.

Article 100

Coordination (Selective)
NEC Page 27

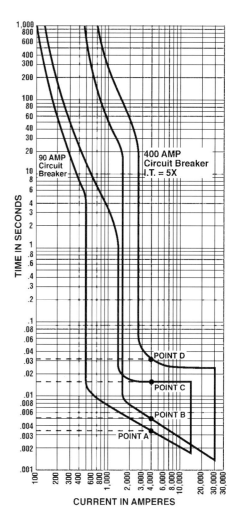

FIGURE 1.5

Article 100

Device
NEC Page 28

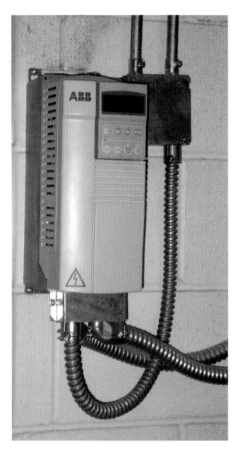

FIGURE 1.6

CHANGE TYPE: Revision (Proposal 1-78; Comment 1-63)

CHANGE SUMMARY: The word *control* has been added to the existing definition of *device* to clarify that a device may also be used to "control" electrical circuits or equipment but not use electric energy.

2005 CODE: "**Device.** A unit of an electrical system that is intended to carry <u>or control</u> but not utilize electric energy."

CHANGE SIGNIFICANCE: This change indicates that a device sometimes does more than "carry but not utilize electric energy." For example, a motor controller carries current as well as interrupts current by making and breaking contacts, but it is not considered utilization equipment because it does not consume electric energy. (Note that the coil will consume a little energy.)

Lighting relays and ordinary switches are considered control devices. Control devices are used extensively in industrial control equipment for all types of machinery and industrial processes.

Article 100

Dwelling Unit
NEC Page 28

CHANGE TYPE: Revision (Proposal 1-83; Comment 1-67)

CHANGE SUMMARY: The previous definition of *dwelling unit* found in Article 100 has been replaced with the definition from other NFPA codes in the interest of consistency.

2005 CODE: "**Dwelling Unit.** <u>A single unit, providing complete, and independent living facilities for</u> one or more ~~rooms for the use of one or more~~ persons, <u>including permanent provisions for</u> ~~as a housekeeping unit with space for~~ <u>living, sleeping, cooking, and sanitation</u> ~~eating, living, and sleeping, and permanent provisions for cooking and sanitation~~."

CHANGE SIGNIFICANCE: The previous requirement that a dwelling unit be a "housekeeping unit" has been deleted. Provisions for living, sleeping, cooking, and sanitation are now all defined as permanent, whereas the previous definition only required the provisions for cooking and sanitation to be permanent.

FIGURE 1.7

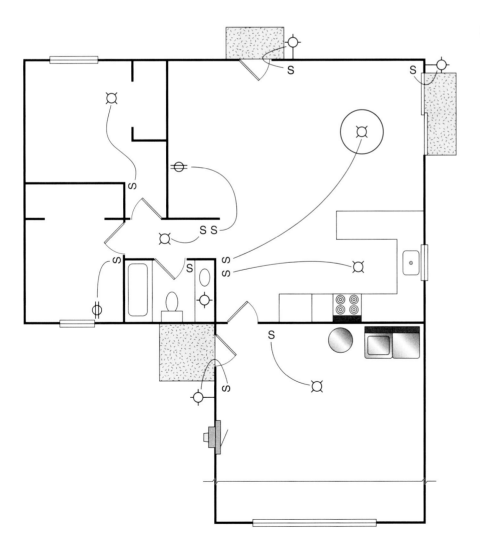

This proposal is a result of the Technical Correlating Committee's request to the Usability Task Group to investigate key definitions in the NEC as they relate to the Building Construction and Safety Code NFPA 5000, Uniform Fire Code NFPA 1, and Life Safety Code NFPA 101. A number of companion proposals to clarify the terms for *guest room*, *guest suite*, and the usage of those terms in the NEC were also included.

The revised definition is that which is found in NFPA 1 and 101 and is being used in the NEC for standardization within the NFPA family of documents. This change is intended to make the definitions consistent without impacting the usage of *dwelling unit* in the NEC.

The change also supports the opinion expressed by CMP-2 that a microwave oven is not considered a permanent cooking appliance and also the philosophy that a "microwave oven does not a kitchen make."

Article 100

Grounded, Solidly
NEC Page 28

CHANGE TYPE: New (Proposal 1-136)

CHANGE SUMMARY: The definition of *grounded, solidly* is now established in Article 100. A similar phrase was used in 230.95 relating to ground-fault protection of equipment.

2005 CODE: "**Grounded, Solidly.** Connected to ground without inserting any resistor or impedance device."

CHANGE SIGNIFICANCE: CMP-1 has added a definition of the term *grounded, solidly* to Article 100 because the term is used in several other articles.

It is recognized that all conductors have resistance or impedance. The application of the term *solidly grounded* suggests that the resistance or impedance of the grounding electrode conductor is ignored. The system is considered "solidly grounded" when it is connected to the grounding electrode with a copper or aluminum grounding electrode conductor, if an impedance device, such as a resistor or inductor, is not located in the path from the service equipment or separately derived system to the grounding electrode.

FIGURE 1.8

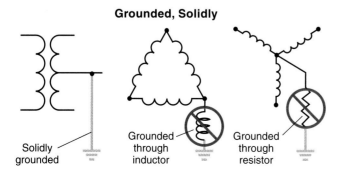

Grounded, Solidly

Solidly grounded

Grounded through inductor

Grounded through resistor

Article 100

Grounding Electrode
NEC Page 29

CHANGE TYPE: New (Proposal 1-97)

CHANGE SUMMARY: This proposal adds a definition of *grounding electrode* to Article 100.

2005 CODE: "**Grounding Electrode.** A device that establishes an electrical connection to the earth."

CHANGE SIGNIFICANCE: The term *grounding electrode* is used in several articles in the NEC. As the definition indicates, grounding electrodes are installed to make a connection to the earth for electrical systems and equipment. This earth connection provides a path for overvoltages, such as results from lightning or when high voltage lines contact lower voltage systems, to flow to the earth. Grounding electrodes also attempt to keep electrical equipment at earth potential. This reduces any shock hazard to people and animals who might come in contact with the equipment while also being in contact with the earth or another grounded object.

The requirement for which grounding electrodes must be used is found in 250.50. Section 250.52 contains the descriptions of grounding electrodes that are permitted for making a connection of the electrical system or equipment to the earth. Rules for installing the grounding electrodes are found in 250.53.

It has been suggested repeatedly that using *device* to define a grounding electrode is improper. We often think of devices in the context of wiring devices such as switches or receptacles. However, the revised definition of the term *device* makes it clear that the term applies more broadly and includes units of electrical systems that carry or control electric energy but are not utilization equipment. While not intended to carry current on a regular basis, grounding electrodes do carry current while dissipating overvoltages into the earth.

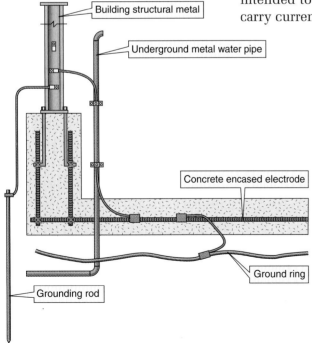

Grounding Electrode

Building structural metal

Underground metal water pipe

Concrete encased electrode

Ground ring

Grounding rod

FIGURE 1.9

CHANGE TYPE: Revised (Proposal 1-98; Comment 1-96)

CHANGE SUMMARY: The definition of *grounding electrode conductor* has been revised to clarify the location where grounding electrode conductors are installed.

2005 CODE: "**Grounding Electrode Conductor.** The conductor used to connect the grounding electrode(s) to the equipment grounding conductor, to the grounded conductor, or to both, at the service, at each building or structure where supplied <u>by a feeder(s) or branch circuit(s),</u> ~~from a common service,~~ or at the source of a separately derived system."

CHANGE SIGNIFICANCE: The deletion of the phrase "from a common service" removes a term that has been confusing to many NEC users. The addition of the phrase "by a feeder(s) or branch circuit(s)" makes the definition technically correct.

By definition, the "service" is supplied from a utility source. The service-disconnecting means separates the system supplied by the utility from the premises wiring. All supply conductors after the service disconnecting means are either feeders or branch circuits. So, any buildings or structures that are separate from the building where the service equipment is located are supplied by either one or more feeders or branch circuits as permitted by 225.30.

Article 100

Grounding Electrode Conductor
NEC Page 29

FIGURE 1.10

Grounding Electrode Conductors

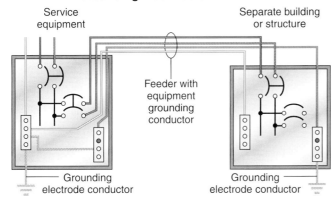

Service equipment

Separate building or structure

Feeder with equipment grounding conductor

Grounding electrode conductor

Grounding electrode conductor

Article 100

Guest Room
NEC Page 29

CHANGE TYPE: New (Proposal 1-101; Comment 1-97)

CHANGE SUMMARY: A definition of *guest room* has been added to Article 100 to ensure a standardized application of NEC rules.

2005 CODE: "**Guest Room.** An accommodation combining living, sleeping, sanitary, and storage facilities within a compartment."

CHANGE SIGNIFICANCE: This definition came from NFPA 101 and is part of the Technical Correlating Committee's ongoing efforts to standardize the use of terms in the NFPA family of codes and standards. *Guest room* is included in several articles, including 210, 220, and 240. It is intended to define a compartment usually consisting of two rooms—one room for living, sleeping, and storage, and another room or area defined as a bathroom. *Guest room* differs from the newly defined *guest suite*—the latter term usually refers to three definable rooms or areas although, by definition, it is only required to have two definable areas.

FIGURE 1.11

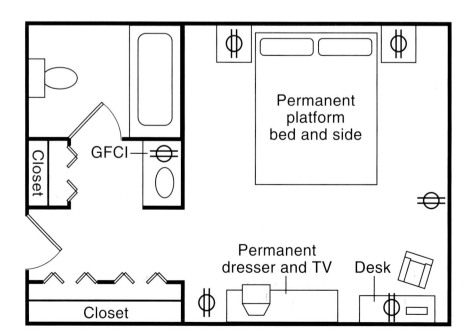

Article 100

Guest Suite
NEC Page 29

CHANGE TYPE: New (Proposal 1-101; Comment 1-97)

CHANGE SUMMARY: A definition of *guest suite* has been added to Article 100 to ensure a standardized application of NEC rules.

2005 CODE: "**Guest Suite.** An accommodation with two or more contiguous rooms comprising a compartment, with or without doors between such rooms, that provides living, sleeping, sanitary, and storage facilities."

CHANGE SIGNIFICANCE: This definition came from NFPA 101 and is also part of the Technical Correlating Committee's ongoing efforts to standardize the use of terms in the NFPA family of codes and standards. The term *guest suite* in used in several articles including 210 and 240. It is intended to define a compartment usually consisting of three rooms or areas, including one room for living, another for sleeping and storage, and a third room or area defined as a bathroom. *Guest suite* differs from the newly defined *guest room*—the former usually consists of two definable rooms or areas in the compartment, although it is only required to have one definable compartment or area.

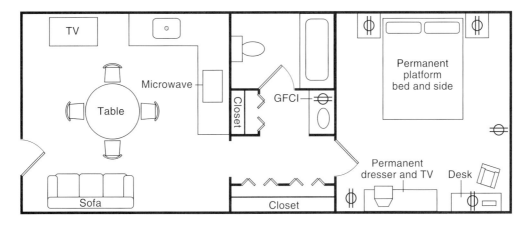

FIGURE 1.12

Article 100

Handhole Enclosure

NEC Page 29

FIGURE 1.13

CHANGE TYPE: New (Proposal 1-109)

CHANGE SUMMARY: A definition of *handhole enclosure* has been added to Article 100. Requirements for the installation of handholes and related conduits and conductors have been added in several other locations in the Code.

2005 CODE: "**Handhole Enclosure.** An enclosure identified for use in underground systems, provided with an open or closed bottom, and sized to allow personnel to reach into, but not enter, for the purpose of installing, operating, or maintaining equipment or wiring or both."

CHANGE SIGNIFICANCE: The addition of this definition and the requirements for installing handholes addresses the widespread industry practice of using underground but flush-mounted handholes for the distribution of underground branch-circuit wiring in conduit and cable systems. Handholes are often used in conjunction with underground wiring methods along streets and in parks and recreational areas for installing landscape lighting and light poles, as well as for other applications where an above-ground box would pose a physical hazard. Handholes are preferable to listed wet-location pull or junction boxes because these boxes are not designed for immersion during prolonged flooding conditions experienced in many parts of the country. A handhole allows for natural drainage through an open bottom.

It is important to note that the handhole is considered to be in a wet location where it is installed at or below the surface of the earth. (See the definition of *location, wet* in Article 100.) As a result, all connections must be made so that the connectors are suitable for the environment. Some twist-on wire connectors are listed for wet locations, as are pressure or compression connectors sealed in heavy-wall heat shrink tubing that are listed for wet locations or direct burial. Other wet location insulating methods include molds and self-hardening mastic.

Other locations in the NEC where requirements or information regarding handholes are being added include:

- 300.15(L), in which a box is not required for wiring in handholes;
- the title and scope of Article 314;
- 314.29, in which wiring in handholes is required to be accessible but is permitted to be covered by gravel, light aggregate, or non-cohesive, granulated soil if its location is effectively identified and accessible for excavation; and
- 314.30 (new), in which the majority of requirements for installing handholes are included.

Article 100

Qualified Person
NEC Page 30

CHANGE TYPE: New (Proposal 1-130; Comment 1-159)

CHANGE SUMMARY: A new FPN has been added following the definition of *qualified person* to indicate that NFPA 70E-2004 contains information on requirements for safety training.

2005 CODE: "**Qualified Person.** One who has skills and knowledge related to the construction and operation of the electrical equipment and installations and has received safety training on the hazards involved.

FPN: Refer to NFPA 70E-2004, *Standards for Electrical Safety in the Workplace*, for electrical safety training requirements."

CHANGE SIGNIFICANCE: Three proposals, all of which were not accepted for various reasons, were made to revise the previous definition of *qualified person*. One proposal asked for an FPN to direct the Code user to NFPA 70E for additional information on requirements for training necessary for the qualified person. Two other proposals asked that the training indicated in the definition of *qualified person* be documented, to allow the authority having jurisdiction to verify that the training being received was appropriate for the work being performed.

One of the International Brotherhood of Electrical Workers (IBEW) proposals asked for the following changes to the definition:

"**Qualified Person.** One who has skills and knowledge related to the construction and operation of the electrical equipment and installations and has received <u>documented</u> safety training on the hazards involved. <u>This training shall ensure that persons shall also be familiar with the proper use of special precautionary techniques, personal protective equipment, insulating and shielding materials, and insulated tools and test equipment when working on or near exposed conductors and or circuit parts that are or can become energized.</u>"

FIGURE 1.14

Several comments were submitted on the proposed changes to the definition. One comment pointed out that the NFPA Glossary of Terms includes five definitions of the term *qualified person*. Two of the definitions include a requirement for "possession of a recognized degree, certificate, professional standing, or skill, and who by knowledge, training, and experience has DEMONSTRATED the ability to deal with problems associated to the subject matter, the work, or the project."

There is a critical role played by qualified person(s) in the NEC. A search of the 2002 NEC shows 91 examples where the term *qualified persons* is used. There are several instances where the provisions of the NEC take into account the value of having qualified personnel service an installation. In exchange for ensuring that only qualified personnel are involved, a somewhat relaxed requirement is often provided for in the NEC.

Therefore, in keeping with the purpose of the NEC as the practical safeguarding of persons and property from the hazards arising from the use of electricity as stated in 90.1(A), ensuring that qualified persons receive the training necessary to safely perform their duties is paramount.

Article 100

Separately Derived System

NEC Page 31

CHANGE TYPE: Revision (Proposal 1-132; Comment 1-167)

CHANGE SUMMARY: Changes were made to the definition of *separately derived system* in order to remove the existing list of typical separately derived systems.

2005 CODE: "**Separately Derived System.** A premises wiring system whose power is derived from a ~~battery, from a solar photovoltaic system, or from a generator, transformer, or converter windings,~~ source of electric energy or equipment other than a service. Such systems have ~~and that has~~ no direct electrical connection, including a solidly connected grounded circuit conductor, to supply conductors originating in another system."

CHANGE SIGNIFICANCE: CMP-1 deleted the list of power sources for separately derived systems rather than add fuel cells as another example of a separately derived system. The revised definition now makes it clear that a separately derived system can be from "equipment." The addition of this term clarifies that a transformer can serve as the source of a separately derived system even though it is not the energy source. In actuality, the service is not the source of electric energy either; the generating station where the electric energy has its origin is really the source for the system provided by the electric utility. On-site power production sources such as generators are obviously also considered an energy source. Fuel cells and solar-photovoltaic systems produce electrical energy as well.

The important requirement that "no direct electrical connection, including a solidly connected, grounded circuit conductor, to supply conductors that originate in another system" remains the determining factor as to whether a system is considered "separately derived." Specific rules for grounding and bonding separately derived systems are found in 250.30.

Generator-Type Separately Derived Alternating Current Systems

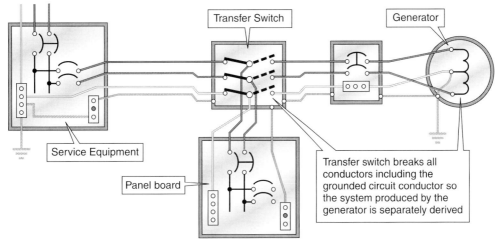

FIGURE 1.15

Article 100

Supplementary Overcurrent Protective Device

NEC Page 31

FIGURE 1.16

CHANGE TYPE: New (Proposal 1-138)

CHANGE SUMMARY: A definition of *supplementary overcurrent protective device* has been added, even though the term, in its precise form, is not used in the Code.

2005 CODE: "**Supplementary Overcurrent Protective Device.** A device intended to provide limited overcurrent protection for specific applications and utilization equipment such as luminaries (lighting fixtures) and appliances. This limited protection is in addition to the protection provided in the required branch circuit by the branch circuit overcurrent protective device."

CHANGE SIGNIFICANCE: Although the precise term is not used in the Code, several variations of *supplementary overcurrent protective device* are used. The most common usage is "supplementary overcurrent protection."

Section 240.10 provides specific requirements on the use of supplementary overcurrent protection devices. As indicated in the definition, supplementary overcurrent protection devices are not permitted to be used as a substitute for branch-circuit overcurrent devices or in place of the branch-circuit protection. Supplementary overcurrent protective devices are not general use devices, as are branch-circuit devices, and must be evaluated for appropriate application in every instance where they are used. They are extremely application-oriented; prior to applying the devices, their differences and limitations must be investigated and found acceptable.

One example of these differences and limitations is that a supplementary overcurrent protective device may have spacings, creepage, a nd clearance that are considerably less than that of a branch-circuit overcurrent protective device. Another example is that branch-circuit overcurrent protective devices have standard overload characteristics to protect branch-circuit and feeder conductors. Supplementary overcurrent protective devices do not have these same standard overload characteristics. Also, supplementary overcurrent protective devices have interrupting ratings that can range from 32 amps to 100,000 amps. When supplementary overcurrent protective devices are considered for proper use, it is important to be sure that the device's interrupting rating equals or exceeds the available short-circuit current and that the device has the proper voltage rating for the installation (including compliance with slash voltage rating requirements, if applicable).

Supplementary overcurrent protective devices are referenced in, but not limited to, the following sections: 240.5, 240.10, 240.24, 422.11, 424.19, 424.22, 424.72, 430.72, 520.50, 690.4, 690.9, and 702.6.

110.12

Mechanical Execution of Work
NEC Page 34

CHANGE TYPE: New (Proposal 1-158)

CHANGE SUMMARY: An FPN has been added to provide information on accepted industry practices and workmanship.

2005 CODE: "FPN: Accepted industry practices are described in ANSI/NECA 1-2000, Standard Practices for Good Workmanship in Electrical Contracting, and other ANSI-approved installation standards."

CHANGE SIGNIFICANCE: As presently written, 110.12 is an undefined performance requirement. While the opening paragraph requires "Electrical equipment to be installed in a neat and workmanlike manner," nothing in other sections of the NEC describes the meaning of "neat and workmanlike." No guidance is given on how to determine when an installation meets this requirement. As a result, determination of compliance is subject to widely varying interpretations, and subsections (A), (B), and (C) describe only a few of the important aspects of "neat and workmanlike" electrical installations.

This new FPN indicates where additional information can be found to assist the installer in making sure that an installation is "neat and workmanlike." It should be emphasized, however, that the FPN is not enforceable. The ANSI standard is likewise unenforceable unless the local jurisdiction adopts it as a minimum requirement. The NECA/ANSI standard becomes the minimum requirement if an architect or electrical engineer includes the standard as the minimum level of performance in the specifications for a project.

FIGURE 1.17

American National Standard

NEiS

NECA 1–2000

Standard practices for
**Good Workmanship
in Electrical Contracting**

Published by
National Electrical Contractors Association **NECA**

110.15

High-Leg Marking
NEC Page 35

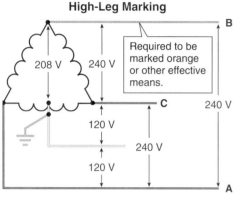

FIGURE 1.18

CHANGE TYPE: Revision (Proposal 1-170)

CHANGE SUMMARY: Changes have been made to this section to broaden the requirements for high-leg marking and to make minor editorial changes.

2005 CODE: "**High Leg Marking.** On a 4-wire, delta-connected system where the midpoint of one phase winding is grounded ~~to supply lighting and similar loads~~, only the conductor or busbar having the higher phase voltage to ground shall be durably and permanently marked ~~either~~ by an outer finish that is orange in color or by other effective means. Such identification shall be placed at each point on the system where a connection is made if the grounded conductor is also present."

CHANGE SIGNIFICANCE: As a result of deleting the phrase, "to supply lighting and similar loads," the application of this rule is expanded to cover all installations of these delta-connected systems where the grounded conductor is present.

The described systems are commonly 240-volt transformers that are delta connected with the midpoint of one transformer center-tapped and grounded. This results in a 120/240-volt, 3-phase, 4-wire system. These systems are commonly installed in smaller commercial installations where it is desired to have 240-volt, 3-phase power for motor loads and 120 volts for receptacle and lighting branch circuits. The voltage to ground from the A and C phases is 120 volts; it is approximately 208 volts from the B phase to ground and to the grounded conductor, referred to as "the conductor or busbar having the higher phase voltage to ground." It is important that the conductor with the higher voltage to ground be identified where the grounded conductor is present so that the conductors are not connected improperly. Equipment intended to operate at 120 volts could be damaged or destroyed if connected on a circuit with 208 volts.

The use of the color orange is not reserved within the NEC for identification of the conductor with the higher voltage to ground. The non-heating leads of electric space-heating cables that operate at 480 volts are also required to be marked in orange, as is one of the isolated circuit conductors for isolated power systems in health care facilities.

CHANGE TYPE: Revision (Proposal 1-176)

CHANGE SUMMARY: This revision adds "meter socket enclosures" to the existing list of equipment requiring flash protection warning.

2005 CODE: "Switchboards, panelboards, industrial control panels, <u>meter socket enclosures,</u> and motor control centers that are in other than dwelling occupancies and are likely to require examination, adjustment, servicing, or maintenance while energized shall be field marked to warn qualified persons of potential electric arc flash hazards. The marking shall be located so as to be clearly visible to qualified persons before examination, adjustment, servicing, or maintenance of the equipment."

CHANGE SIGNIFICANCE: The intent of this section in the NEC is to provide a warning for both qualified and unqualified persons of the arc flash dangers present in the area of pieces of equipment that "are likely to require examination, adjustment, servicing, or maintenance while energized." As can be readily seen, this phrase is identical to that in 110.26(A) to determine when working space is required around this equipment. It is therefore safe to conclude that where working space is required in 110.26(A), flash protection warning is required for the equipment identified in 110.16.

   Previously, 110.16 allowed the installation of a multi-gang meter enclosure containing service conductors outside of a commercial strip store without an arc flash warning, while requiring such a warning for the panelboards that serve as the service equipment.

   The addition of "meter socket enclosures" is necessary due to the frequent addition and/or removal of meters "while energized" in com-

110.16

Flash Protection
NEC Page 35

FIGURE 1.19

mercial strip stores as well as similar occupancies. These enclosures are often multi-gang enclosures and are most often worked on while energized to install or remove meters or conductors as tenants move in and out of occupancies, as well as for upgrades to existing occupancies. These meter socket enclosures represent the highest level of incident energy available in the system, second only to the terminals on the secondary of the transformer.

In some cases, meter socket enclosures contain "service conductors" that are protected only by the primary overcurrent device on the supply side of the transformer. These utility-owned transformers are not protected at 125% of the primary current of the transformer as would be required by the general rule in 450.3(B). The overcurrent devices on the primary of these transformers are set at a higher level since the utilities are not bound by the NEC and use higher values of protection on the primary side of the transformer to limit nuisance outages. This represents an extremely high value of incident energy when compared to the downstream equipment, which will have overcurrent protection as per the NEC and is included in the scope of 110.16. (Note that although the meter socket enclosures may contain service conductors, they fall under the scope of the NEC. The meter socket enclosures are "premises wiring," as they either are on the load side of the "service point" or are the "service point.") The addition of "meter socket enclosures" to the scope of 110.16 is practical and safety driven—the incident energy is at the greatest value.

CHANGE TYPE: Revision (Proposals 1-207a and 1-208)

CHANGE SUMMARY: Changes have been made to the conditions used for the application of working spaces for electrical equipment given in Table 110.26(A)(1). The depth of working space for Condition 2 for a voltage of 151 to 600 has been changed from 1 meter to 1.1 meters ($3^1/_2$ feet).

2005 CODE: "Condition 1 — Exposed live parts on one side <u>of the working space</u> and no live or grounded parts on the other side of the working space, or exposed live parts on both sides <u>of the working space</u> <u>that are</u> effectively guarded by ~~suitable wood or other~~ insulating materials. ~~Insulated wire or insulated busbars operating at not over 300 volts to ground shall not be considered live parts.~~

Condition 2 — Exposed live parts on one side <u>of the working space</u> and grounded parts on the other side. Concrete, brick, or tile walls shall be considered as grounded.

Condition 3 — Exposed live parts on both sides of the working space ~~(not guarded as provided in Condition 1) with the operator between~~."

CHANGE SIGNIFICANCE: While these changes are referred to in the proposal as "editorial," in reality, the changes are significant. The phrase "of the working space" (or "of the work space") is now consistent in each condition.

The sentence "Insulated wire or insulated busbars operating at not over 300 volts to ground shall not be considered live parts" has been deleted from Condition 1 as it is not considered necessary based upon the definition of the term *Live Parts* in Article 100. *Live Parts* is defined as "energized conductive components." This deletion may cause some interpretation problems with the authority having jurisdiction as some may now apply the working space rules to enclosures having conductors that are energized but insulated for the voltage applied. It is important to remember that these working space rules apply where equipment is "likely to require examination, adjustment, servicing, or maintenance while energized." Usually, connections are not made to energized conductors in pull boxes, wireways, and auxiliary gutters, so the working space rules do not apply.

Working space can be described as a sphere that has dimensions not less than the width of the equipment and not less than 30 in. wide, is the height of the equipment but not less than $6^1/_2$ ft, and is not less than the depth required in Table 110.16(A)(1). It is sometimes helpful to draw a rectangle representing the working space on the floor plan of the equipment room or area.

It should also be understood that the working spaces around multiple pieces of electrical equipment are permitted to overlap. Generally, one piece of electrical equipment cannot infringe on the working space of another. According to Section 110.26(A)(3), "other equipment that is associated with the electrical installation and is located above or

110.26 (A)(1)
Depth of Working Space
NEC Page 36

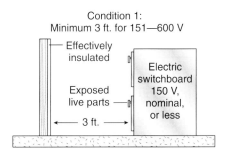

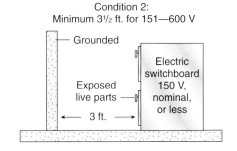

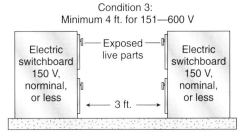

FIGURE 1.20

below the electrical equipment shall be permitted to extend not more than 150 mm (6 in.) beyond the front of the electrical equipment." However, it should be understood that this "intrusion" only relates to electrical equipment on the same side of the working space and not to equipment that is across from the working space.

CHANGE TYPE: Revision (Proposal 1-217; Comment 1-248)

CHANGE SUMMARY: The phrase "and over 6 ft wide" has been deleted from the description of "large equipment."

2005 CODE: "(2) Large Equipment. For equipment rated 1200 amperes or more ~~and over 1.8 m (6 ft) wide~~ that contains overcurrent devices, switching devices, or control devices, there shall be one entrance to the required working space not less than 610 mm (24 in.) wide and 2.0 m (6¹/₂ ft) high at each end of the working space. Where the entrance has a personnel door(s), the door(s) shall open in the direction of egress and be equipped with panic bars, pressure plates, or other devices that are normally latched but open under simple pressure."

CHANGE SIGNIFICANCE: The substantiation for the deletion of the 6-foot rule as a part of the condition for determining when equipment is considered "large" mentions that the hazard from an arc blast is directly related to the electrical rating and not to the physical size of equipment. In reality, the arc flash and arc blast are related to the available

110.26 (C)(2)
Large Equipment
NEC Page 37

FIGURE 1.21

incident energy at a location, which is based on the voltage, available short-circuit current, separation between the electrodes (phases), distance a worker's body parts are from the arcing fault, and duration of the fault. Also, it has been reported that some equipment with a rating of 1200 amperes or more may be less than 6 feet in length but still pose an unacceptable risk to a worker who may be trapped in a narrow work space if there is a blowup.

Where it is determined that equipment is "large," the working space must comply with one of three requirements:

1. An entrance to the working space must be provided at each end.

2. A continuous and unobstructed way of exit travel from the working space must be provided.

3. The working space from Table 110.26(A)(1) must be doubled.

CHANGE TYPE: Relocated (Proposal 1-240)

CHANGE SUMMARY: The action on this proposal relocates the existing Part IV of Article 314 to the new Part V of Article 110, with a title of "Manholes and Other Electric Enclosures Intended for Personnel Entry, All Voltages."

2005 CODE: The text of this relocated part has not changed significantly from the 2002 NEC.

CHANGE SIGNIFICANCE: This proposal originated with the Technical Correlating Committee to move the existing Part IV of Article 314 to the new Part V of Article 110. It was considered part of the plan to improve the organization of the NEC by locating similar requirements together.

   An important part of the rules for manholes is that they be "sufficient in size" and have adequate working space around equipment likely to require examination, adjustment, servicing, or maintenance while energized.

Article 110, Part V

Manholes and Other Electric Enclosures Intended for Personnel Entry, All Voltages
NEC Page 41

FIGURE 1.22

CHAPTER 2

Wiring and Protection, Articles 200–285

200.6(B)

Sizes Larger Than 6 AWG

NEC Page 43

CHANGE TYPE: Revision (Proposal 5-16)

CHANGE SUMMARY: The means of identification of grounded conductors larger than 6 AWG has been revised to allow gray as well as white markings at terminations.

2005 CODE: "**(B) Sizes Larger Than 6 AWG.** An insulated grounded conductor larger than 6 AWG shall be identified <u>by one of the following means:</u> ~~either~~

(1) By a continuous white or gray outer finish. ~~or~~

(2) By three continuous white stripes ~~on other than green insulation~~ along its entire length <u>on other than green insulation.</u> ~~or~~

(3) At the time of installation by a distinctive white <u>or gray</u> marking at its terminations. This marking shall encircle the conductor or insulation."

CHANGE SIGNIFICANCE: Since the Code previously recognized both white and gray colors to identify the grounded conductor, and since previous panel statements and Code text recognize that white and gray are different colors, the widespread practice of identifying grounded conductors of some systems with white or gray marking or tape should also be recognized and legitimized.

The Code Panel does not stipulate which colors should be used for which system. However, if a conductor with white insulation or a non-white conductor with white tape is permitted, and a conductor with gray insulation is permitted, then a conductor with gray phase tape should also be permitted. Allowing the use of gray-taped conductors will facilitate the consistent application of color-coding of systems by designers or installers.

Conductors larger than 6 AWG are most commonly provided with black insulation, so it is important to allow both white and gray markings on these conductors to indicate grounded conductors.

FIGURE 2.1

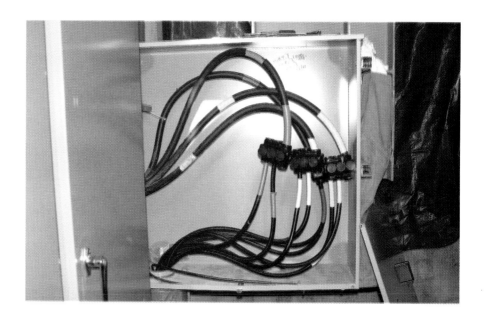

CHANGE TYPE: Revision (Proposal 5-20)

CHANGE SUMMARY: Changes to 200.6(D) require that where grounded conductors of different systems are installed in common raceways, cable, boxes, etc, each system is to be identified and the means of identification is to be permanently posted at each branch-circuit panelboard.

2005 CODE: "**(D) Grounded Conductors of Different Systems.** Where grounded conductors of different systems are installed in the same raceway, cable, box, auxiliary gutter, or other type of enclosure, each grounded conductor shall be identified by system. Identification that distinguishes each system grounded conductor shall be permitted by one of the following means:

(1) One system grounded conductor, if required, shall have an outer covering conforming to 200.6(A) or 200.6(B).

(2) The grounded conductor(s) of other systems shall have a different outer covering conforming to 200.6(A), 200.6(B) or by an outer covering of white or gray with a readily distinguishable, colored stripe other than green running along the insulation.

(3) Other and different means of identification as allowed by 200.6(A) or (B) that will distinguish each system grounded conductor.

This means of identification shall be permanently posted at each branch-circuit panelboard."

CHANGE SIGNIFICANCE: For many years the use of the colors gray and white were not permitted as a means to distinguish grounded conductors of different systems. With the removal of the term "natural gray" in the 2002 NEC, gray is now a color permitted to identify a grounded conductor. With the changes to 200.6 in the 2002 NEC, both gray and white are now permitted.

The Code Panel action on the proposal makes the following significant changes. First, when grounded conductors of different systems are in the same raceway, cable, box, auxiliary gutter, or other type of enclosure, they are required to be distinguished from each other. Secondly, the proposal clarifies the permitted means to distinguish one system grounded conductor from another. Finally, it is noted that the means of identification is required to be permanently posted at each branch-circuit panelboard. This posting requirement is identical to that for branch circuits in 210.5(C).

200.6(D)
Grounded Conductors of Different Systems
NEC Page 43

FIGURE 2.2

An Example of Conductor Color Coding

	120/240-V,1 Ph	208Y/120-V,3-Ph	480Y/277-V,3-Ph
Phase A	Black	Black	Brown
Phase B	Red	Red	Orange
Phase C		Blue	Yellow
Neutraql	White	White with red Stripe	Gray

Articles 210 & 220

Guest Rooms and Guest Suites

CHANGE TYPE: New (Proposal 2-5)

CHANGE SUMMARY: The term *guest room* has been replaced with the phrase "guest rooms or guest suites" in 210.6(A), 210.60 title, 210.60(A), 210.70(B) title and text, 220.3(B)(8), 220.3(B)(10), and 220.12(B).

2005 CODE: See NEC for actual wording.

CHANGE SIGNIFICANCE: This proposal was submitted by the NEC Technical Correlating Committee and is intended to make the use of terms uniform in the NEC, NFPA 5000 (the NFPA Building Code), NFPA 1 (the NFPA Fire Protection Code), and NFPA 101 (the Life Safety Code). This proposal was accompanied by a number of proposals intended to create consistent definitions and usage in the NEC. A proposal to define *guest room* and *guest suites* in Article 100 was submitted by the TCC as well.

Many facilities today are constructed using the "suite" concept; these changes will provide guidance as to the requirements to be applied. For NEC purposes, guest rooms and guest suites are considered identical whether the accommodation consists of one or more rooms.

The definitions accepted by CMP 1 are as follows:

"**Guest Room.** An accommodation combining living, sleeping, sanitary, and storage facilities within a compartment."

"**Guest Suite.** An accommodation with two or more contiguous rooms comprising a compartment, with or without doors between such rooms, that provides living, sleeping, sanitary, and storage facilities."

FIGURE 2.3

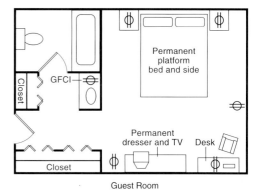

Guest Room

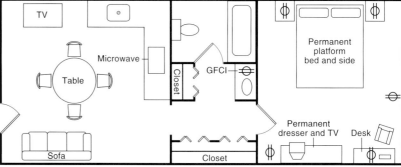

Guest Suite

CHANGE TYPE: Revision (Proposals 2-12 and 2-30; Comment 2-16)

CHANGE SUMMARY: Changes have been made to subdivisions (A) and (B) while subdivision (D) shown below, has been deleted because the new 210.5(C) applies to all ungrounded conductors, not just multiwire branch circuits. These changes expand the coverage of this section to other than dwelling units and to branch circuits that originate in other than panelboards, such as switchboards.

2005 CODE: "**(A) General.** Branch circuits recognized by this article shall be permitted as multiwire circuits. A multiwire circuit shall be permitted to be considered as multiple circuits. All conductors shall originate from the same panelboard <u>or similar distribution equipment</u>." FPN: A 3-phase, 4-wire, wye-connected power system used to supply power to non-linear loads may necessitate that the power system design allow for the possibility of high harmonic neutral currents.

210.4 (A), (B), and (D)

Multiwire Branch Circuits
NEC Page 46

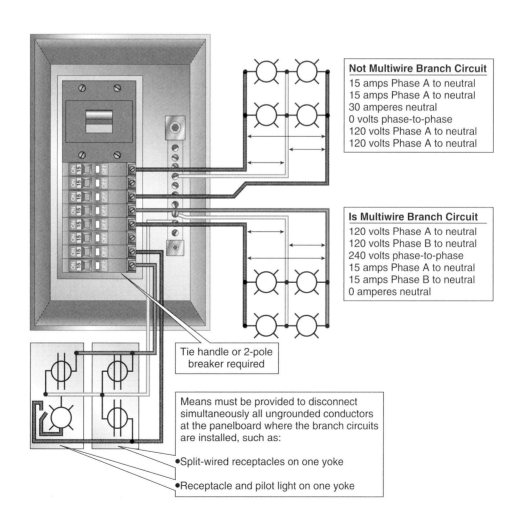

Not Multiwire Branch Circuit
15 amps Phase A to neutral
15 amps Phase A to neutral
30 amperes neutral
0 volts phase-to-phase
120 volts Phase A to neutral
120 volts Phase A to neutral

Is Multiwire Branch Circuit
120 volts Phase A to neutral
120 volts Phase B to neutral
240 volts phase-to-phase
15 amps Phase A to neutral
15 amps Phase B to neutral
0 amperes neutral

Tie handle or 2-pole breaker required

Means must be provided to disconnect simultaneously all ungrounded conductors at the panelboard where the branch circuits are installed, such as:

•Split-wired receptacles on one yoke

•Receptacle and pilot light on one yoke

FIGURE 2.4

(B) Devices or Equipment ~~Dwelling Units~~. Where ~~In dwelling units,~~ a multiwire branch circuit <u>supplies</u> ~~supplying~~ more than one device or equipment on the same yoke, <u>a means</u> shall be provided ~~with a means~~ to disconnect simultaneously all ungrounded conductors <u>supplying those devices or equipment</u> at the <u>point</u> ~~panelboard~~ where the branch circuit <u>originates</u> ~~originated~~.

(C) Line-to-Neutral Loads. Multiwire branch circuits shall supply only line-to-neutral loads.

Exception No. 1: A multiwire branch circuit that supplies only one utilization equipment.

Exception No. 2: Where all ungrounded conductors of the multiwire branch circuit are opened simultaneously by the branch-circuit overcurrent device.

FPN: See 300.13(B) for continuity of grounded conductor on multiwire circuits.

(D) ~~**Identification of Ungrounded Conductors.** Where more than one nominal voltage system exists in a building, each ungrounded conductor of a multiwire branch circuit, where accessible, shall be identified by phase and system. This means of identification shall be permitted to be by separate color coding, marking tape, tagging, or other approved means and shall be permanently posted at each branch circuit panelboard."~~

CHANGE SIGNIFICANCE: While the substantiation for the proposal intended to primarily address the identification rules for multiwire branch-circuit conductors, several other significant changes have been made to this section.

By adding the phrase "or similar distribution equipment" to the opening paragraph, the rule now clearly applies wherever branch circuits originate. In addition to panelboards, branch circuits originate at motor control centers, switchboards, and industrial control panels. The sentence "All conductors shall originate from the same panelboard or similar distribution equipment" is intended to prohibit the practice of making a multiwire branch circuit by picking up a conductor from phase A and B from one piece of distribution equipment and phase C and the neutral from another piece of distribution equipment.

The changes made to (B) broaden the requirement to all occupancies for providing a disconnection means where the multiwire branch circuit supplies more than one device or equipment on the same wiring device yoke. Common applications of this rule include combination wiring devices, such as receptacle/switch, receptacle/pilot light, or receptacle/receptacle, where the tab between duplex receptacles is removed. Note that the definition of "receptacle" in Article 100 includes the statements, "A single receptacle is a single contact device with no other contact device on the same yoke. A multiple receptacle is two or more contact devices on the same yoke." So a duplex receptacle with the tab between receptacles on the ungrounded (hot) side that is supplied by a multiwire branch circuit requires the simultaneous disconnecting means.

This rule no longer applies to only dwelling units. The most common way of complying with this rule is to install a common trip circuit breaker or single pole circuit breakers with identified tie handles in the panelboard or other distribution equipment where the multiwire branch circuit supplying the wiring device originates.

Former 210.4(D), "Identification of Ungrounded Conductors," has been deleted and its requirements have been added to 210.5(C). By making this change, the rules on identification of ungrounded conductors apply to all types of branch circuits, not only to multiwire branch circuits.

210.5(C)

Ungrounded Conductors

NEC Page 46

CHANGE TYPE: New (Proposal 2-30)

CHANGE SUMMARY: This new subdivision requires identification of all branch-circuit conductors where the conductors of the premises wiring system are supplied by more than one nominal voltage system. The rule formerly applied only to multiwire branch circuits. No specific color-coding is specified for the systems. Posting the means of identification is required. In addition to accepting the new section, CMP-2 has deleted former 210.4(D).

2005 CODE: "**(C) Ungrounded Conductors.** Where the premises wiring system has branch circuits supplied from more than one nominal voltage system, each ungrounded conductor of a branch circuit, where accessible, shall be identified by system. The means of identification shall be permitted to be by separate color coding, marking tape, tagging, or other approved means and shall be permanently posted at each branch-circuit panelboard or similar branch-circuit distribution equipment."

CHANGE SIGNIFICANCE: The proposal that resulted in this change was an effort to reenergize consideration of the need for identification of branch-circuit conductors based upon the color of the insulation.

Widespread industry practice, including job-site installation practices, engineering specifications, and job specifications, dictates that a means of branch-circuit and feeder identification, based on the color of conductor insulation, be implemented for a project. A table included with the proposal suggested conductor insulation colors based upon widespread industry practice, which requires black, red, and blue for lower voltages and brown, orange, and yellow for higher voltages.

FIGURE 2.5

An Example of Conductor Color Coding			
	120/240-V,1 Ph	208Y/120-V,3-Ph	480Y/277-V,3-Ph
Phase A	Black	Black	Brown
Phase B	Red	Red	Orange
Phase C		Blue	Yellow
Neutral	White	White with red Stripe	Gray

However, the Code Panel chose to not require branch-circuit identification based only on the color of conductor insulation. CMP-2 states in its Panel Statement on the proposal, "The panel agrees that the identification of the ungrounded conductors for branch circuits is needed. However, the panel notes that color code is only one of several acceptable means to accomplish this." The Panel therefore allows the identification to be by "separate color coding, marking tape, tagging, or other approved means." While the "other approved means" is not specified, the authority having jurisdiction can accept means such as number or letter tape labels for identification of the branch circuits.

The language of the new section is not new to the Code. It was brought over from previous 210.4(D), where it applied to only multiwire branch circuits. By locating the rule in 210.5(C), it applies to installations where the premise's wiring system has branch circuits from more than one nominal voltage system. The term *nominal voltage* is defined in Article 100 as, "A nominal value assigned to a circuit or system for the purpose of conveniently designating its voltage class (e.g., 120/240 volts, 480Y/277 volts, 600 volts)." So, if a building has a service at 480Y/277 volts and a transformer creates a 208Y/120 volt system, two nominal voltages exist in the building and the rule in 210.5(C) applies. If the building is supplied by a 208Y/120 volt service and no other nominal voltage systems are present in the building, the new rule does not apply.

Likewise, if a building is supplied by a 208Y/120 volt service and a 208Y/120 volt generator is connected through a transfer switch, the new rule does not apply since there is only one nominal voltage system in the building.

Finally, the rule now applies to not only panelboards where branch circuits originate but to "similar branch-circuit distribution equipment." This includes switchboards, motor control centers, and industrial control equipment. The means of identification used in the building or structure must be "permanently posted" at each branch-circuit panelboard or similar branch-circuit distribution equipment.

While this rule applies to branch circuits only, a nearly identical requirement has been added to 215.12 that requires identification of feeders if there are feeders supplied from more than one nominal voltage system.

210.8(A) (7)

GFCI Protection, Dwelling Units

NEC Page 47

Laundry Room

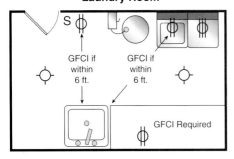

FIGURE 2.6

CHANGE TYPE: Revision (Proposal 2-42; Comment 2-38)

CHANGE SUMMARY: GFCI protection is now required for 125-volt, single-phase, 15- and 20-ampere receptacles installed within 6 feet of laundry and utility sinks.

2005 CODE: "(7) <u>Laundry, utility, and</u> wet bar sinks—where the receptacles are installed ~~to serve the countertop surfaces and are located~~ within 1.8 m (6 ft) of the outside edge of the sink."

CHANGE SIGNIFICANCE: The obvious change to this section is the addition of laundry and utility sinks to the locations where GFCI protection for personnel is required in dwelling units. However, neither of these terms is defined. The location of the laundry sink should be fairly obvious, as the Code has required for a number of years that there be a laundry circuit installed for laundry equipment. The laundry sink is usually located in the laundry room or laundry area.

No exception to the requirement for GFCI protection is provided for a receptacle that is located specifically to serve an appliance such as a 120-volt washing machine or clothes dryer. The only qualification for whether GFCI protection is required is whether the receptacle is within 6 feet of the outside edge of the sink.

A similar expanded requirement now applies to wet bar sinks as well. Previously, GFCI protection was required only if the receptacle was installed to serve countertop surfaces and was within 6 feet from the edge of the sink. As now written, GFCI protection is required for *any* receptacle installed within 6 feet of the outside edge of the sink, even though it may be installed to serve another appliance or for another purpose. Though not clear in the Code, this revised rule may be interpreted by the authority having jurisdiction as "Any receptacle that can be contacted by a 6-foot cord requires GFCI protection."

The term *utility sink* has not been used previously in the Code. Under a fairly broad interpretation, a utility sink can be any sink other than a kitchen sink, a laundry sink, or a basin located in the bathroom.

210.8(B)(2)

GFCI Protection, Other Than Dwelling Units

NEC Page 47

CHANGE TYPE: Revision (Proposal 2-85; Comment 2-58)

CHANGE SUMMARY: The qualifier "commercial and institutional" has been added to this section, along with a description of *commercial and industrial kitchen*. GFCI protection is only required for commercial and institutional kitchens in 210.8(B)(2).

2005 CODE: "(2) <u>Commercial and institutional</u> kitchens. <u>For the purposes of this section, a kitchen is an area with a sink and permanent facilities for food preparation and cooking</u>."

CHANGE SIGNIFICANCE: Several proposals were presented to clarify the application of the rule on GFCI protection in other than dwelling-unit kitchens. The expansion of GFCI requirements to include non-dwelling unit kitchens in the 2002 NEC was not uniformly enforced due to conflicting interpretations as to what constitutes a kitchen.

Building codes, including NFPA 5000, do not define a kitchen. This proposal intended to require GFCI protection in employee break rooms that contain the defined features, not just in commercial (restaurant) kitchens. It was stated that employee break rooms with a sink and a microwave oven are areas where older ungrounded equipment, including toasters and coffee makers, are likely to be used.

However, the Code Panel stopped short of expanding the GFCI requirements to employee break rooms—unless they meet the new definition of *kitchen*. In its statement, the Panel said, "There are many different designs and configurations of commercial kitchens. Certainly it is reasonable to conclude that the kitchen is an area where there is a sink and provisions for food preparation and sanitation. This definition distinguishes commercial and institutional kitchens from those

FIGURE 2.7

areas that might have a portable cooking appliance." Clearly, the appliances used in the commercial or institutional kitchen are required to be "permanent facilities for food preparation and cooking." Note that a sink is another qualifier to determine a "commercial or institutional kitchen" but a refrigerator is not required.

There continue to be no exceptions to the requirement for GFCI protection in commercial or institutional kitchens for applicable 125-volt receptacles. Several proposals were made for the 2005 NEC to exempt various receptacles from the GFCI requirement. All were rejected by the Code Panel.

CHANGE TYPE: New (Proposal 2-70)

CHANGE SUMMARY: The Code now requires GFCI protection for 15- and 20-ampere, 125-volt, single-phase receptacles for other than dwelling units in outdoor spaces in public areas.

2005 CODE: "(4) Outdoors in public spaces—for the purpose of this section, a public space is defined as any space that is for use by, or is accessible to, the public."

CHANGE SIGNIFICANCE: There were several proposals to expand the requirements for GFCI protection to all of the areas required for dwelling units. Once again, the Code Panel did not expand the requirements for GFCI protection to all the locations required.

This proposal for expansion of GFCI requirements originated with the U.S. Consumer Product Safety Commission (CPSC). The CPSC estimates that there were 170 accidental electrocutions associated with consumer products in 1999 in the United States, the latest year available.

Based on information provided on death certificates, the CPSC identified six product categories involved in significant numbers of electrocutions: power tools and equipment, installed wiring, antenna products, large appliances, small appliances, and lighting products. The CPSC conducted follow-up investigations of a selection of the electrocution deaths that occurred over the seven-year period 1994–2000 in the United States to find causal factors. In total, 209 incidents were documented with sufficient detail (many of these investigations included on-site visits by CPSC representatives, photographs, investigation reports by local authorities, and interviews with people with

210.8(B)(4)

GFCI Protection, Other Than Dwelling Units
NEC Page 47

FIGURE 2.8

relevant knowledge). From these in-depth investigation reports, conditions that led to death were noted, and practical solutions to reduce the risk of electrocution under similar circumstances in the future became evident.

While most consumer electrocutions occur in and around the home, electrocutions frequently are reported in public and community settings associated with cord-connected power equipment and large appliances, such as coin-operated machines, tools, pumps, and cleaning equipment. Providing GFCI protection for receptacle outlets, both indoor and outdoor, located in public and community access areas will address these high-risk locations.

The following is a partial list of electrocutions that occurred at areas covered by this proposal.

- Williston, ND; October 8, 1996: A 9-year-old male was electrocuted at an indoor recreation center by a cord- and plug-connected coin-operated machine.
- Waco, TX; May 29, 1997: A 44-year-old male was electrocuted on public property while servicing a cold-drink vending machine.
- Clanton, AL; August 21, 1995: A 10-year-old male was electrocuted at a motel when he came in contact with a vending machine.
- Melbourne, FL; June 24, 1998: A 19-year-old male was electrocuted in a rented unit in a public storage facility when he came in contact with an electric guitar and microphone plugged into a receptacle outlet not properly grounded.
- Tallahassee, FL; November 11, 1988: A 26-year-old male was electrocuted after he contacted a change machine and a snack machine simultaneously in a college student lounge.
- Corwith, IA; May 4, 2000: A 16-year-old male was electrocuted when he contacted a power tool plugged into a receptacle outlet within a shelter in a municipal park.

Having reviewed the CPSC investigations, the Panel revised the recommendation to apply to outdoor public locations, but did not accept the application of the requirement to *all* locations accessible to the public. The Panel concluded that the substantiation provided clear direction to apply this to outdoor locations because of damp or wet environments. Several of the incidents described in the substantiation involved equipment installed outdoors in locations accessible to the general public and supplied from 125-volt, 15- or 20-ampere receptacles.

This revision recognizes that the rule does not apply to those industrial and other locations where the general public does not have access.

CHANGE TYPE: New (Proposal 2-99)

CHANGE SUMMARY: This change expands the requirement for GFCI protection of receptacles installed for servicing HVAC equipment. The GFCI rule no longer applies *only* to rooftop locations.

2005 CODE: "(5) Outdoors, where installed to comply with 210.63."

CHANGE SIGNIFICANCE: Several Code cycles ago, a requirement was added to 210.63 for a receptacle to be installed within 25 feet of all heating, air-conditioning, and refrigeration equipment. This receptacle aids personnel who service this equipment in inclement weather. Much of the equipment they service is grounded, which can present a ground-fault return path from any faulty tool being held by an individual.

Since the requirement is known to be specific to personnel use and it is clear that the outdoor environment is a location where ground-fault protection is needed to protect personnel using electrical equipment, this proposal was accepted and 210.8(B)(5) was modified accordingly.

210.8(B)(5)

GFCI Protection, Other Than Dwelling Units
NEC Page 47

FIGURE 2.9

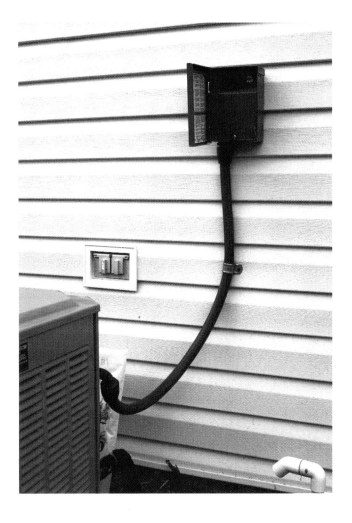

210.8(C)

Boat Hoists

NEC Page 47

CHANGE TYPE: New (Proposal 2-47; Comment 2-65)

CHANGE SUMMARY: GFCI protection is now required for boat hoists installed at dwelling unit locations, whether connected through a cord and plug or "hard wired."

2005 CODE: "**(C) Boat Hoists.** Ground-fault circuit-interrupter protection for personnel shall be provided for outlets that supply boat hoists installed in dwelling unit locations and supplied by 125-volt, 15- and 20-ampere branch circuits."

CHANGE SIGNIFICANCE: This proposal was one of several originating with the Consumer Product Safety Commission (CPSC). In the 1980s, in cooperation with manufacturers of boat hoist equipment, CPSC staff identified motor-operated boat hoist equipment intended for use at residential settings as consumer products that needed GFCI protection, to reduce the risk of electrocution when using this equipment while near bodies of water. (This action was taken in response to a number of electrocutions with boat hoists in residential settings where the equipment did not have GFCI protection.) During the processing of the 1996 NEC, the CPSC presented a proposal to have boat hoists covered by GFCI requirements. As stated at that time (Proposal 2-82), there had been at least three electrocutions over a three-year period from boat hoists.

FIGURE 2.10

The CPSC stated that grounding provisions associated with fixed wiring cannot be relied upon alone for adequate electrocution protection for boat hoists. Associated dangers include the fact that these installations are exposed to harsh weather conditions, the presence of moisture corrosive to the typical boat hoist metallic apparatus, and the presence of cords used with the motor and motor control wiring harnesses commonly found on fixed-wired, electrically powered boat hoists.

Including the requirement for GFCI protection for boat hoists at dwelling units harmonizes the NEC with accepted manufacturing practice. The addition also will reduce confusion and the chance that products without GFCI protection will enter the market in the future.

210.12 (B) and Exception

Arc-Fault Circuit-Interrupter Protection in Dwelling Unit Bedrooms

NEC Page 48

CHANGE TYPE: Revised and new exception (Proposal 2-134a; Comment 2-87a)

CHANGE SUMMARY: (B) has been revised to require that the AFCI protection be of the combination type, but permits the present branch/feeder type to be used until January 1, 2008. A new exception permits the AFCI to be installed at other than the origination of the branch circuit under the specified conditions.

2005 CODE: "**210.12 Arc-Fault Circuit-Interrupter Protection.**
(B) Dwelling Unit Bedrooms. All ~~branch circuits that supply 125~~ 120-volt, single-phase, 15- and 20-ampere <u>branch circuits supplying</u> outlets installed in dwelling unit bedrooms shall be protected by ~~an~~ <u>a</u> <u>listed</u> arc-fault circuit interrupter, <u>combination type,</u> <u>installed</u> ~~listed~~ to provide protection of the ~~entire~~ branch circuit.

<u>Branch/Feeder AFCIs shall be permitted to be used to meet the requirements of 210.12(B) until January 1, 2008.</u>

<u>**FPN:** For information on types of arc-fault circuit interrupters, see UL 1699-1999, Standard for Arc-Fault Circuit Interrupters.</u>
<u>*Exception: The location of the arc-fault circuit interrupter shall be permitted to be at other than the origination of the branch circuit in compliance with (a) and (b):*</u>
<u>*(a) The arc-fault circuit interrupter shall be installed within 1.8 m (6 ft) of the branch circuit overcurrent device as measured along the branch circuit conductors.*</u>
<u>*(b) The circuit conductors between the branch circuit overcurrent device and the arc-fault circuit interrupter shall be installed in a metal raceway or a cable with a metallic sheath.*</u>"

FIGURE 2.11

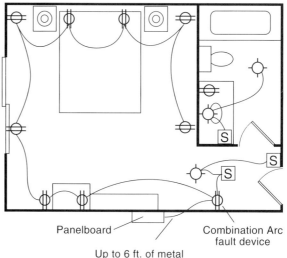

Panelboard

Combination Arc fault device

Up to 6 ft. of metal raceway or metal-clad cable permitted

CHANGE SIGNIFICANCE: Several proposals and comments were presented both for expansion as well as elimination of AFCI requirements. As in previous NEC cycles, the merits of AFCI technology were debated extensively by CMP-2. At the end of the process for the 2005 NEC, there is no expansion of usage required nor are there any exceptions for outlets installed in dwelling unit bedrooms, such as for smoke alarms.

As noted above, the combination type AFCI will be required effective January 1, 2008. This is the approximate date on which the 2008 NEC will be available. Combination-type AFCIs are available in receptacle devices. This lead time gives manufacturers of wiring device-type AFCIs to have their products fully developed and in the marketplace.

The Panel statement on the comment reads, "The panel recognizes the level of safety provided by AFCIs; however, since the combination-type technology is relatively new, a transition period has been established for the industry to meet this requirement."

The Underwriters Laboratories Web site identifies six types of AFCIs. The branch/feeder type is in present production and is available only as circuit breakers. Four different manufacturers are identified as having listed products in this category. Two manufacturers are identified as having listed combination-type AFCIs.

The Underwriters Laboratories Guide Card provides the following information about branch/feeder and combination-type AFCIs:

Arc Fault Circuit Interrupters, Branch/Feeder Type—A device intended to be installed at the origin of a branch circuit or feeder, such as at a panelboard. It is intended to provide protection of the branch-circuit wiring, feeder wiring, or both, against unwanted effects of arcing. This device also provides limited protection to branch-circuit extension wiring. It may be a circuit-breaker-type device or a device in its own enclosure mounted at or near a panelboard.

Arc Fault Circuit Interrupters, Combination Type—An AFCI that complies with the requirements for both branch/feeder and outlet circuit AFCIs. It is intended to protect downstream branch-circuit wiring, cord sets, and power-supply cords.

Other types of AFCIs include outlet branch circuit type, outlet circuit type, cord type, and portable type.

210.18

Guest Rooms and Guest Suites

NEC Page 49

CHANGE TYPE: New (Proposal 2-242; Comment 2-139)

CHANGE SUMMARY: A new section has been added to ensure that branch circuits that meet the rules for dwelling units are installed when guest room or guest suites are provided with permanent provisions for cooking.

2005 CODE: "**210.18. Guest Rooms and Guest Suites.** Guest rooms and guest suites that are provided with permanent provisions for cooking shall have branch circuits and outlets installed to meet the rules for dwelling units."

CHANGE SIGNIFICANCE: This new section is intended to plug a hole that would have otherwise been created if guest rooms or guest suites in lodging facilities had permanent cooking appliances but were not required to have small appliance branch circuits (as required of other dwelling units).

With this change, if guest rooms or guest suites meet the definition of a dwelling unit by having permanent provisions for cooking, the units will be required to have all branch circuits installed as though they were dwelling units. This new section is intended to coordinate with the new definitions of the terms *guest room* and *guest suite* in Article 100 as well as the requirements in 210.60 for installation of receptacle outlets.

FIGURE 2.12

Guest Suite with permanent provisions for cooking

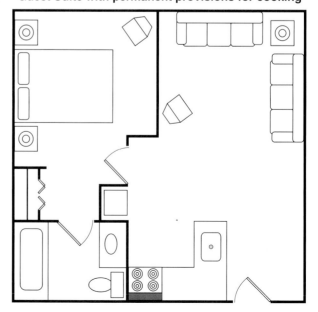

CHANGE TYPE: Revision (Proposal 2-206)

CHANGE SUMMARY: Changes to this section clarify that the two small-appliance branch circuits required by 210.11(C)(1) must serve all the receptacle outlets in the specified rooms, including wall, floor, and countertop receptacles.

2005 CODE: "**(1) Receptacle Outlets Served.** In the kitchen, pantry, breakfast room, dining room, or similar area of a dwelling unit, the two or more 20-ampere small-appliance branch circuits required by 210.11(C)(1) shall serve all <u>wall and floor</u> receptacle outlets covered by 210.52(A), <u>all countertop outlets covered by</u> ~~and~~ 210.52(C), and receptacle outlets for refrigeration equipment."

CHANGE SIGNIFICANCE: The revised wording, though appearing to be editorial, clarifies the application of the requirements, which adds to the user-friendliness of the Code. Though floor outlets are rarely used in kitchens, they may be used in dining or breakfast rooms and now must clearly be supplied by one of the two 20-ampere small-appliance branch circuits.

The substantiation for the proposal claimed that the previous wording of 210.52(B)(1) was self-conflicting. The current wording starts by stating which receptacle outlets this section is referring to ("in the kitchen, pantry, breakfast room, dining room, or similar area of a dwelling unit") but later in the same section paragraph requires that all receptacle outlets covered by 210.52(A) and (C) be served by the two or more 20-ampere small-appliance branch circuits required by 210.11(C)(1). These two statements are conflicting, as 210.52(A) refers to receptacles in every kitchen, family room, dining room, living room, parlor, library, den, sunroom, bedroom, recreation room, or similar room or area of dwelling units.

This revision resolves the debate over the intent and meaning of the section, both in the field between contractor and inspector, and in court between attorneys and judges.

210.52 (B)(1)

Small Appliances, Receptacle Outlets Served
NEC Page 52

FIGURE 2.13

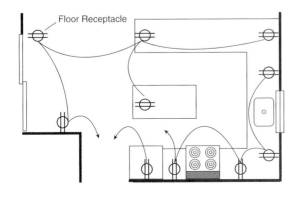
Floor Receptacle

210.52 (C)(1) Exception and (2)

Countertops, Wall Counter Spaces, and Island Counter Spaces
NEC Page 52

CHANGE TYPE: New (Proposal 2-211a; Comment 2-125)

CHANGE SUMMARY: A new exception and figure are being added to 210.52(C)(1) to clarify that receptacle outlets are not required on a wall directly behind a range or sink in the installation described in the figure. Also, language similar to the exception for 210.52(C)(1) is being added to (C)(2) to clarify when a range top or sink divides an island counter space into two spaces. If divided into two spaces, a receptacle outlet is required for each island counter space.

2005 CODE: "**(C) Countertops.** In kitchens and dining rooms of dwelling units, receptacle outlets for counter spaces shall be installed in accordance with 210.52(C)(1) through (C)(5).

(1) **Wall Counter Spaces.** A receptacle outlet shall be installed at each wall counter space that is 300 mm (12 in.) or wider. Receptacle outlets shall be installed so that no point along the wall line is more than 600 mm (24 in.) measured horizontally from a receptacle outlet in that space.

FIGURE 2.14

Sink or range extending from face of counter

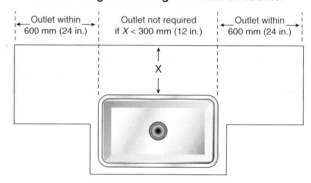

Sink or range mounted in corner

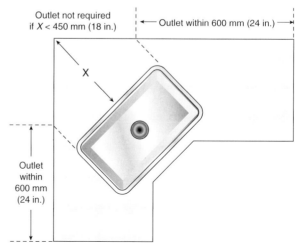

Exception: Receptacle outlets shall not be required on a wall directly behind a range or sink in the installation described in Figure 210.52.

(2) Island Counter Spaces. At least one receptacle shall be installed at each island counter space with a long dimension of 600 mm (24 in.) or greater and a short dimension of 300 mm (12 in.) or greater. Where a rangetop or sink is installed in an island counter and the width of the counter behind the rangetop or sink is less than 12 inches, the rangetop or sink is considered to divide the island into two separate countertop spaces as defined in 210.52(C)(4)."

CHANGE SIGNIFICANCE: Very little substantiation was provided by the Code Panel for adding the exception and new figure. However, both will solve many interpretation problems installers experience in the field. It clarifies the issue as to whether a receptacle outlet is required in corner spaces and counter spaces that are occupied by a sink or range. It answers the question, "Is the space behind the sink or range considered wall space so far as requiring receptacles is concerned?"

The figure now also illustrates where the measurement begins for kitchen sinks that are mounted in the corner. As seen in the figure, a straight line is drawn from the edge of the sink back to the wall. That is the point where the 24-inch measurement begins for determining wall-mounted receptacles required for kitchen counter spaces.

The new sentence that is added to 210.52(C)(2) for island counter spaces will aid and support a uniform application and interpretation of the Code as to when the island counter is divided into two spaces. Where the installation of a range top or sink divides the island counter into two spaces, a receptacle outlet is required for each space. This is emphasized in 210.52(C)(4).

Peninsular counter spaces in 210.52(C)(3) are beyond the coverage of these new rules as the requirements apply to wall receptacles and island counter spaces. It would be wise to apply the new rules for the island countertop to peninsular countertops as well.

210.52 (D) and Exception

Dwelling Unit Bathrooms

NEC Page 53

CHANGE TYPE: New Exception (Proposal 2-229)

CHANGE SUMMARY: This change clarifies the requirement for a wall receptacle in the bathroom by adding a new exception permitting a receptacle on the face of the basin cabinet.

2005 CODE: "**(D) Bathrooms.** In dwelling units, at least one ~~wall~~ receptacle outlet shall be installed in bathrooms within 900 mm (3 ft) of the outside edge of each basin. The receptacle outlet shall be located on a wall or partition that is adjacent to the basin or basin countertop.

Exception: The receptacle shall not be required to be mounted in the wall or partition where it is installed on the side or face of the basin cabinet not more than 300 mm (12 in.) below the countertop."

CHANGE SIGNIFICANCE: This change makes the location of the receptacle required to serve the basin area in dwelling unit bathrooms more flexible. While the receptacle is still required to be within 3 feet of the outside edge of each basin, it can now be mounted on the face of the basin cabinet not more than 12 inches below the countertop.

The substantiation uses as an example a problem an installer had with the previous Code language. On a master bath renovation, the custom-built vanity had a solid granite top with mirrors on the left, front, and right sides that went from the countertop to the ceiling. There was no place to put the required receptacle outlet "on the wall."

This proposal indicates the Code Panel's responsiveness to real-world situations and also demonstrates the value of soliciting Code proposals from those who actually use the NEC.

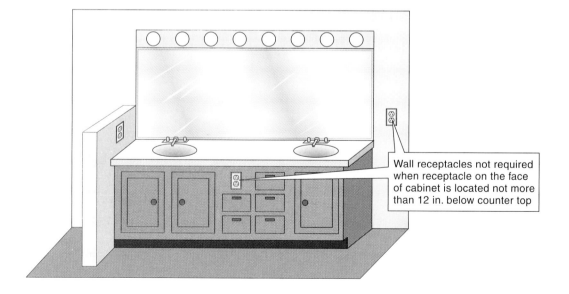

Wall receptacles not required when receptacle on the face of cabinet is located not more than 12 in. below counter top

FIGURE 2.15

CHANGE TYPE: Revision (Proposal 2-135; Comment 2-136)

CHANGE SUMMARY: The new second paragraph in this section requires one outdoor receptacle for certain multifamily dwelling units.

2005 CODE: "**(E) Outdoor Outlets.** For a one-family dwelling and each unit of a two-family dwelling that is at grade level, at least one receptacle outlet accessible at grade level and not more than 2.0 m ($6^1/_2$ ft) above grade shall be installed at the front and back of the dwelling. ~~See 210.8(A)(3).~~

For each dwelling unit of a multifamily dwelling where the dwelling unit is located at grade level and provided with individual exterior entrance/egress, at least one receptacle outlet accessible from grade level and not more than 2.0 m ($6^1/_2$ ft) above grade shall be installed. See 210.8(A)(3)."

CHANGE SIGNIFICANCE: For the first time requirements for outdoor receptacles for certain multifamily dwellings are added. An outdoor receptacle outlet will be required for each dwelling unit of a multifamily dwelling for those dwelling units that are located at grade level and that have an individual exterior entrance or exit. Unlike one- and two-family dwellings, the outdoor receptacle is required to be accessible from grade level and not more than $6^1/_2$ feet above grade.

Interpretation of 210.52(E) from NFPA staff and from members of the Code Panel on the previous rule has been that a receptacle outlet located above a porch or deck that is accessible at grade does not meet the requirement of being "accessible from grade level." To qualify, the receptacle must be accessible while the person is at grade level and must not be more than $6^1/_2$ feet above grade.

210.52(E)

Dwelling Unit Outdoor Receptacle Outlets
NEC Page 53

FIGURE 2.16

210.60 (A)

Guest Rooms or Guest Suites, General

NEC Page 54

CHANGE TYPE: Revision (Proposal 2-242; Comment 2-139)

CHANGE SUMMARY: This proposal originated with the Technical Correlating Committee and is intended to correlate with NFPA 5000, NFPA 1, and NFPA 101.

2005 CODE: "**(A) General.** Guest rooms <u>or guest suites</u> in hotels, motels, and similar occupancies shall have receptacle outlets installed in accordance with 210.52(A) and 210.52(D). Guest rooms <u>or guest suites provided with permanent provisions for cooking</u> ~~meeting the definition of a dwelling unit~~ shall have receptacle outlets installed in accordance with all of the applicable rules in 210.52."

CHANGE SIGNIFICANCE: This proposal was a result of the Technical Correlating Committee's request that the usability task group investigate key definitions in the NEC as they relate to National Building Code NFPA 5000, Fire Prevention Code NFPA 1, and Life Safety Code NFPA 101. This proposal was accompanied by a number of proposals to use a consistent definition for *dwelling unit* and to clarify the terms *guest room*, *guest suite*, and the usage of those terms in the NEC.

It is clear that CMP-2 intends for guest rooms (and suites) of hotels and motels that have all the same provisions as a dwelling unit to have

FIGURE 2.17

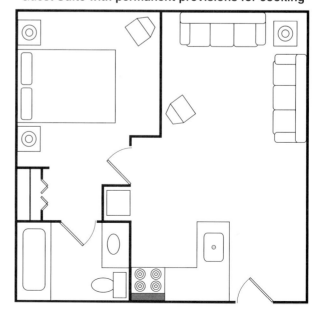

Guest Suite with permanent provisions for cooking

receptacle outlets installed according to 210.52. This proposal retains that same thinking, but utilizes the new definitions of *guest room* and *guest suite* to make that point. It is important to note that the added language "provided with permanent provisions for cooking" is key because the basic definition does not include such provisions.

The existing language has been modified to eliminate confusion between the electrical and building codes as to the precise definition of these types of accommodations. The purpose of this section is to state which wiring rules are to be used without attempting to modify the actual occupancy definition that may have an unintended effect on any other code.

In addition, CMP-2 has reaffirmed its position that permanent provisions for cooking do not include equipment such as a microwave oven. It should also be noted that new 210.18 requires guest rooms and guest suites that have permanent provisions for cooking to have all of the branch circuits required for dwelling units. This would include receptacle and lighting circuits.

210.63 Exception

Heating, Air-Conditioning, and Refrigeration Equipment Outlet

NEC Page 54

CHANGE TYPE: New Exception (Proposal 2-250)

CHANGE SUMMARY: A new exception has been added to this section to explain that a receptacle outlet is not required for servicing evaporative coolers for one- and two-family dwellings.

2005 CODE: "**210.63 Heating, Air-Conditioning, and Refrigeration Equipment Outlet.** A 125-volt, single-phase, 15- or 20-ampere-rated receptacle outlet shall be installed at an accessible location for the servicing of heating, air-conditioning, and refrigeration equipment. The receptacle shall be located on the same level and within 7.5 m (25 ft) of the heating, air-conditioning, and refrigeration equipment. The receptacle outlet shall not be connected to the load side of the equipment disconnecting means.

Exception: A receptacle outlet shall not be required at one- and two-family dwellings for the service of evaporative coolers."

CHANGE SIGNIFICANCE: The individual who submitted this proposal described the steps needed to install, service, and replace evaporative coolers at dwellings. When evaporative coolers are installed, the first step is to cut a hole in the roof for the ductwork. This is the only part of the installation where 120-volt power is required. Currently, many installers use battery-powered saws. Service work on a cooler only involves replacing the motor, pads, or pump. No 120-volt powered tools are required for servicing the equipment. Years later, when the cooler is replaced, the procedures are similar. The old cooler is disconnected and removed from the roof. The new one is secured, most likely with battery-powered tools.

The Code Panel concluded that the type of equipment covered by the new exception does not have the same service and maintenance requirements as do other types of air-conditioning equipment. Thus a GFCI-protected receptacle outlet at the same level of the evaporative cooler is not required.

FIGURE 2.18

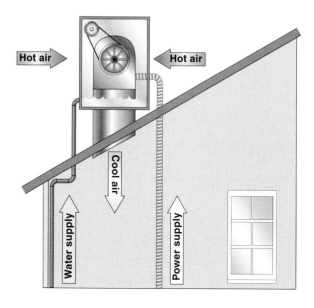

CHANGE TYPE: Revision and New Exceptions (Proposal 2-264)

CHANGE SUMMARY: This is another Technical Correlating Committee proposal, which is related to usability. "Guest Suites" has been added to the title and text. Other changes have made requirements for lighting outlets closer to those for dwelling units. Two exceptions have also been added, similar to the exceptions for 210.70(A)(1) for dwelling units.

2005 CODE: "**(B) Guest Rooms or Guest Suites.** In hotels, motels, or similar occupancies, guest rooms or guest suites shall have at least one wall-switch-controlled lighting outlet ~~or wall-switch-controlled receptacle shall be~~ installed in every habitable room and bathroom. ~~in guest rooms in hotels, motels, or similar occupancies~~.
    Exception No. 1: In other than bathrooms and kitchens where provided, one or more receptacles controlled by a wall switch shall be permitted in lieu of lighting outlets.
    Exception No. 2: Lighting outlets shall be permitted to be controlled by occupancy sensors that are (1) in addition to wall switches or (2) located at a customary wall switch location and equipped with a manual override that will allow the sensor to function as a wall switch."

CHANGE SIGNIFICANCE: This is another proposal from the Technical Correlating Committee to bring the requirements for guest rooms and suites up to date.

    As a result of the review by the usability task group of definitions of *dwelling unit*, *guest room*, and *guest suite*, it was realized that there was an inconsistency in 210.70(B) with respect to lighting outlets in guest rooms and suites of hotels and motels.

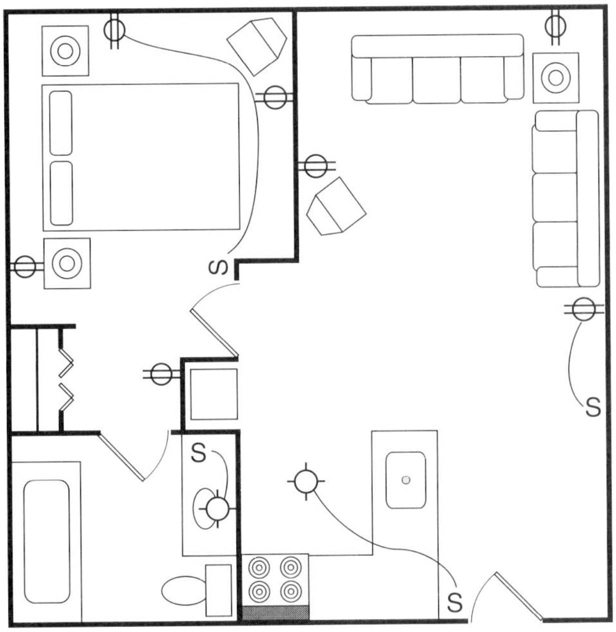

210.70 (B)

Lighting Outlets Required, Guest Rooms or Guest Suites
NEC Page 54

FIGURE 2.19

The previous text could be interpreted to allow a switch-controlled receptacle to be used in lieu of a lighting outlet in a bathroom.

This revision uses similar language from 210.70(A) to make it clear that where these types of accommodations consist of more than one room, each room is to be provided with a lighting outlet. This is consistent with the addition of the term *guest suite* to the NEC.

Additionally, guest rooms typically consist of a sleeping room and a separate bathroom; this revision clarifies that within the bathroom a wall-switch-controlled receptacle is not permitted to serve in lieu of a lighting outlet. In those units where a kitchen is provided, wall-switched-controlled receptacles are also not permitted to serve in lieu of a lighting outlet.

The editorial rearrangement of the main paragraph text was necessary to make sure that the installation requirement applied to guest rooms and guest suites.

CHANGE TYPE: New (Proposal 2-270; Comment 2-148)

CHANGE SUMMARY: The first paragraph of this section has been revised, a new paragraph has been added on the minimum size of the feeder grounded conductor, the previous language on the minimum size of a conductor based upon the number of circuits has been deleted, and other editorial changes have been made.

2005 CODE: "**(1) General.** Feeder conductors shall have an ampacity not less than required to supply the load as <u>calculated</u> ~~computed~~ in Parts ~~II~~, III, ~~and~~ IV, <u>and V</u> of Article 220. The minimum feeder-circuit conductor size, before the application of any adjustment or correction factors, shall have an allowable ampacity not less than the non-continuous load plus 125 percent of the continuous load.
   Exception: Where the assembly, including the overcurrent devices protecting the feeder(s), is listed for operation at 100 percent of its rating, the allowable ampacity of the feeder conductors shall be permitted to be not less than the sum of the continuous load plus the non-continuous load.
   <u>The size of the feeder circuit grounded conductor shall not be smaller than that required by 250.122, except that 250.122(F) shall not apply where grounded conductors are run in parallel.</u>
   Additional minimum sizes shall be as specified in <u>215.2(A)(2) and (A)(3)</u> ~~(2), (3), and (4)~~ under the conditions stipulated.
   ~~(2) For Specified Circuits. The ampacity of feeder conductors shall not be less than 30 amperes where the load supplied consists of any of the following number and types of circuits:~~
   ~~(1) Two or more 2-wire branch circuits supplied by a 2-wire feeder~~
   ~~(2) More than two 2-wire branch circuits supplied by a 3-wire feeder~~

215.2 (A)(1)

Minimum Rating and Size, Feeders Not More Than 600 Volts, General
NEC Page 55

FIGURE 2.20

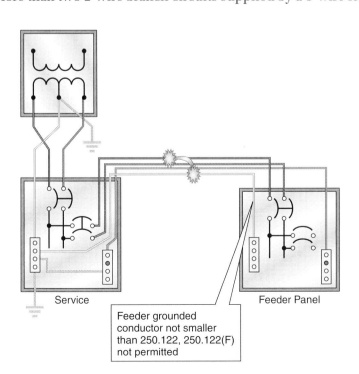

Service   Feeder Panel

Feeder grounded conductor not smaller than 250.122, 250.122(F) not permitted

~~(3) Two or more 3-wire branch circuits supplied by a 3-wire feeder~~
~~(4) Two or more 4-wire branch circuits supplied by a 3-phase, 4-wire feeder~~

(2) ~~(3)~~ Ampacity Relative to Service-~~Entrance~~ Conductors. The feeder conductor ampacity shall not be less than that of the service-~~entrance~~ conductors where the feeder conductors carry the total load supplied by service-~~entrance~~ conductors with an ampacity of 55 amperes or less."

CHANGE SIGNIFICANCE:　The substantiation for this proposal cites a situation where load calculations from Article 220 may require 500-kcmil phase conductors but only a 14 AWG grounded (neutral) conductor. If the overcurrent device protecting the 500-kcmil conductors is a 350-amp breaker, and the neutral and a phase short circuited, the 14 AWG grounded (neutral) conductor may have too much impedance and not allow the breaker to trip fast enough, causing destruction of the 14 AWG grounded (neutral) conductor.

The neutral needs to perform like an equipment grounding conductor in this case. According to Table 250.122, the minimum size grounded conductor for this circuit would be a 3 AWG rather than the 14 AWG.

The phrase "except that 250.122(F) shall not apply where grounded conductors are run in parallel" requires that a grounded conductor not smaller than that given in Table 250.122 be installed in each conduit where the feeder conductors are installed in parallel in separate raceways.

Finally, the grounded conductor must be sized not smaller than required for the calculated load in 220.61.

For feeders over 600 volts in 215.2(B), a new second sentence has been added to read, "Where installed, the size of the feeder circuit grounded conductor shall not be smaller than that required by 250.122, except that 250.122(F) shall not apply where grounded conductors are run in parallel." Identical substantiation to that for feeders not more than 600 volts was provided.

CHANGE TYPE: New (Proposal 2-289; Comment 2-159)

CHANGE SUMMARY: This proposal adds identification requirements for grounded, equipment grounding, and ungrounded conductors. The result is similar to the revised requirements in 210.5(C) for color-coding of branch circuits. In addition, a means of identification is required to be posted when more than one nominal voltage system is supplied.

2005 CODE: "**215.12 Identification for Feeders**
  (A) Grounded Conductor. The grounded conductor of a feeder shall be identified in accordance with 200.6.
  (B) Equipment Grounding Conductor. The equipment grounding conductor shall be identified in accordance with 250.119.
  (C) Ungrounded Conductors. Where the premises wiring system has feeders supplied from more than one nominal voltage system, each ungrounded conductor of a feeder, where accessible, shall be identified by system. The means of identification shall be permitted to be by separate color coding, marking tape, tagging, or other approved means and shall be permanently posted at each feeder panelboard or similar feeder distribution equipment."

CHANGE SIGNIFICANCE: As stated above, the identification of grounded (often neutral) conductors is now required to comply with 200.6 and with 250.119 for equipment grounding conductors.

The requirements for identification of ungrounded feeder conductors are similar to those for branch circuits in 210.5(C). When feeders for the premises' wiring system are supplied from more than one nominal voltage system, each ungrounded conductor of a feeder, where accessible, is required to be identified by system. Examples of different nominal voltage systems include 120/240-volt, 1-phase; 208Y/120-volt, 3-phase; and 480Y/277-volt, 3-phase systems.

As indicated, the means of identification is permitted to be by separate color-coding, marking tape, tagging, or other approved means. A notice of the means of identification used for each system is required to be permanently posted at each feeder panelboard or at similar feeder distribution equipment.

215.12

Identification of Feeders
NEC Page 56

FIGURE 2.21

An Example of Conductor Color Coding

	120/240-V,1 Ph	208Y/120-V,3-Ph	480Y/277-V,3-Ph
Phase A	Black	Black	Brown
Phase B	Red	Red	Orange
Phase C		Blue	Yellow
Neutral	White	White with red Stripe	Gray

Article 220

Branch-Circuit, Feeder, and Service Calculations

CHANGE TYPE: Reorganization (Proposal 2-292)

CHANGE SUMMARY: Article 220 was restructured by a task group of the Technical Correlating Committee and submitted for consideration. The proposal included a new Fine Print Note for 220.1, scope, and Figure 220.1, which shows graphically the organization of Article 220. In addition, the article was renumbered.

2005 CODE: See NEC for actual wording.

2002 Article 220		2005 Article 220	
Part	Sections	Part	Sections
I. General	220.1–4	I. General	220.1–5
II. Feeders and Services	220.10–22	II. Branch Circuit Load Calculations	220.10–18
III. Optional Calculations for Computing Feeder and Service Loads	220.30–36	III. Feeder and Service Service Load Calculations	220.40–61
IV. Method for Computing Farm Loads	220.40–41	IV. Optional Feeder and Service Load Calculations	220.80–88
		V. Farm Load Calculation	220.100–103

CHANGE SIGNIFICANCE: The Code Panel accepted the proposal to reorganize Article 220 and subsequently modified it to include panel actions on several other proposals.

In addition to the reorganization, a new 220.3 and cross reference Table 220.3 were added to provide application of other articles where load calculations are required. (This highlights the fact that not all load

FIGURE 2.22

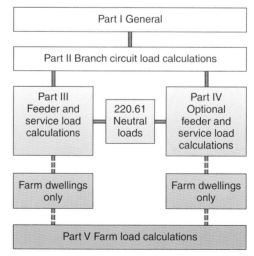

NEC Figure 220.1

calculations are included in Article 220.) Here are a few of the other sections where load calculations are required:

Calculation	Article
Air-Conditioning and Refrigerating Equipment, Branch-Circuit Conductor Sizing	440
Cranes and Hoists, Rating and Size of Conductors	610
Electric Welders, Ampacity Calculations	630
Electrically Driven or Controlled Irrigation Machines	675
Electrolytic Cell Lines	668

A universal change throughout the 2005 NEC is the replacement of "computed load" and "demand load" with "calculated load." For example, the term *demand load* is being replaced with *calculated load* to improve clarity and to reduce confusion, especially when related to "calculated load" and "demand factor." The term *calculated load* includes demand factors, the concepts of diversity, and historical data.

220.82 (C) and Example D2(c)

Dwelling Unit, Heating and Air-Conditioning Load

NEC Pages 68, 718

CHANGE TYPE: Revision (Proposal 2-344)

CHANGE SUMMARY: This revision changes the heating and air-conditioning load subsection of optional load calculations for dwelling units. Specifically, the method of calculating heat pump loads has been revised, which may result in a lower feeder and service load calculation when heat pumps are used.

In addition, Example D2(c) in Annex D has been revised to correlate with the changes in 220.82(C).

2005 CODE: "**(C) Heating and Air-Conditioning Load.** The largest of the following six selections (load in kVA) shall be included:

(1) 100 percent of the nameplate rating(s) of the air conditioning and cooling.

(2) 100 percent of the nameplate rating(s) of the <u>heating when a heat pump is used without any supplemental electric heating</u> ~~heat pump compressors and supplemental heating unless the controller prevents the compressor and supplemental heating from operating at the same time.~~

(3) 100 percent of the nameplate ratings of electric thermal storage and other heating systems where the usual load is expected to be continuous at the full nameplate value. Systems qualifying under this selection shall not be calculated under any other selection in 220.82(C).

(4) <u>100 percent of the nameplate rating(s) of the heat pump compressor and 65 percent of the supplemental electric heating for central electric space heating systems. If the heat pump compressor is prevented from operating at the same time as the supplementary heat, it does not need to be added to the supplementary heat for the total central space heating load.</u> ~~65 percent of the nameplate rating(s) of the central electric space heating, including integral supplemental heating in heat pumps where the controller prevents the compressor and supplemental heating from operating at the same time.~~

(5) 65 percent of the nameplate rating(s) of electric space heating if less than four separately controlled units.

(6) 40 percent of the nameplate rating(s) of electric space heating if four or more separately controlled units."

CHANGE SIGNIFICANCE: The method of determining the calculated load for dwellings with heat pump systems has reverted to the language in the 1996 NEC. Now, 100 percent of the load will be added when the heat pump is used without any supplementary (resistance) heat. If supplementary heat can operate at the same time as the compressor, 100 percent of the heat pump plus 65 percent of the supplementary heat is used in the calculation. This was 100 percent of both loads in the 1999 and 2002 NEC.

Heat pumps typically operate as heating systems in the winter and air-conditioning systems in the summer. Supplemental (strip) resistance heating elements are often added to give a faster recovery from a cooler to a warmer thermostat setting. Strip heaters also function as a backup heat source should the compressor fail for any reason or be turned off when the efficiency falls below an acceptable level in cold weather. Usually, a two-stage thermostat controls the operation of the compressor and supplemental heat. The first stage calls the compressor and the second stage calls the supplemental heat.

The example in Annex D shows a reduction from a 175- to 150-ampere service that includes a heat pump system from the 2002 to the 2005 NEC.

FIGURE 2.23

225.17

Masts as Supports
NEC Page 67

CHANGE TYPE: Revision (Proposal 4-9a; Comment 4-6)

CHANGE SUMMARY: The previous 225.17, "Means of Attachment to Buildings," has become 225.16(B). The requirements added to 225.17 for masts used for outside feeders and branch circuits parallel those for services in 230.28.

2005 CODE: "**225.17 Masts as Supports.** Where a mast is used for the support of final spans of feeders or branch circuits, it shall be of adequate strength or be supported by braces or guys to withstand safely the strain imposed by the overhead drop. Where raceway-type masts are used, all raceway fittings shall be identified for use with masts. Only the feeder or branch-circuit conductors specified within this section shall be permitted to be attached to the feeder and/or branch-circuit mast."

CHANGE SIGNIFICANCE: CMP-4's decision to add requirements for the installation of masts used as the support of overhead spans makes Article 225 mirror the requirements for services in Article 230. The new requirements are somewhat vague; the section uses the phrases "be of adequate strength" and "withstand safely the strain imposed by the overhead drop." Neither of these phrases is specific as to the minimum size conduit required for the various length of overhead span or size of conductors. As a result, the authority having jurisdiction (AHJ) will be responsible for interpreting the requirements added to this section.

Some AHJs have local regulations for the minimum size conduit required for service masts, as well as methods required for securing and bracing a mast. When there is no local AHJ, consult the serving utility for a copy of their regulations, which often include drawings or specifications for service masts. These resources can typically be used for guidance in the absence of specific requirements in the NEC.

FIGURE 2.24

CHANGE TYPE: Revision (Proposal 4-16; Comment 4-20)

CHANGE SUMMARY: Exterior raceways arranged to drain.

2005 CODE: "**225.22 Raceways on Exterior Surfaces of Buildings or Other Structures.** Raceways on <u>exteriors</u> ~~exterior surfaces~~ of buildings or other structures shall be <u>arranged to drain and shall be</u> raintight <u>in wet locations</u> ~~and arranged to drain~~.

Exception: Flexible metal conduit, where permitted in 348.12(1), shall not be required to be raintight."

CHANGE SIGNIFICANCE: The substantiation for the proposal points out that not all locations on exterior surfaces of buildings or structures are wet locations. In damp exterior locations, standard knockout type boxes are not prohibited by 314.15(A), nor are fittings such as couplings and connectors or locknut/bushing terminations required to be raintight.

Changes to this section coordinate with the definition of *wet location* and *damp location* in Article 100. Raceways on the exterior of buildings and structures are required to be "arranged to drain" in both wet and damp locations. No explanation is offered for how this is to be accomplished, such as constructing sloping raceways or providing drain or weep holes. A similar requirement has been in 230.53 for services for many editions of the Code. The NEC Handbook contains the following commentary, "The goal of 230.53 is to prevent water from entering internal electrical equipment through the raceway system. Service raceways exposed to the weather must have raintight fittings and drain holes. During the installation of raceways in masonry, provisions to drain and divert water should be made to prevent the entrance of surface water, rain, or water from poured concrete."

Raintight is defined in Article 100 as "Constructed or protected so that exposure to a beating rain will not result in the entrance of water under specified test conditions." Manufacturers of conduit, cable, and tubing fittings can provide guidance on which fittings are designed or listed to provide a raintight connection.

225.22

Raceways on Exterior Surfaces of Buildings or Other Structures
NEC Page 68

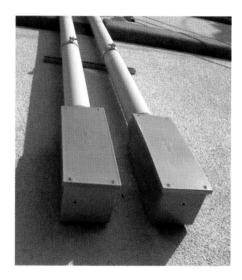

FIGURE 2.25

225.30

Number of Supplies, Special Conditions

NEC Page 69

CHANGE TYPE: Revision (Proposals 4-20a, 4-25, and 4-25a; Comment 4-19)

CHANGE SUMMARY: Editorial changes have clarified the requirement that a building supplied by a feeder or branch circuit shall be supplied by only one of them unless the installation complies with one of the existing conditions. Also, a new subsection (6) has been added to permit another supply for "enhanced reliability."

2005 CODE: "**225.30 Number of Supplies.** Where more than one building or other structure is on the same property and under single management, each additional building or other structure served that is <u>served by a branch circuit or feeder</u> on the load side of the service disconnecting means shall be supplied by one feeder or branch circuit unless permitted in 225.30(A) through (E). For the purpose of this section, a multiwire branch circuit shall be considered a single circuit.

(A) Special Conditions. Additional feeders or branch circuits shall be permitted to supply the following:

(1) Fire pumps

(2) Emergency systems

(3) Legally required standby systems

(4) Optional standby systems

(5) Parallel power production systems

(6) <u>Systems designed for connection to multiple sources of supply for the purpose of enhanced reliability</u>"

CHANGE SIGNIFICANCE: This was a proposal by CMP-4 to address "an ongoing issue for a number of Code cycles." The Code Panel states, "Redundant supplies are often necessary where supplying critical loads,

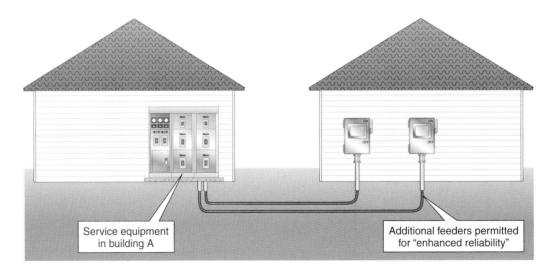

Service equipment in building A

Additional feeders permitted for "enhanced reliability"

FIGURE 2.26

such as medical laboratories that would not be normally covered under Article 517, banks, or computer loads."

Absent from 225.30 is any guidance on what "systems designed for connection to multiple sources of supply" are. Typically, double-ended switchboards with a tie fit this description. When this equipment is installed at the service position, it often is supplied from different utility transformers in order to achieve additional reliability of supply.

Note that this provision is not intended to relate to emergency systems, legally required standby systems, or optional standby systems as those systems are already provided for in (2), (3), and (4) of this section.

230.2(A) (6)

Number of Services, Special Conditions

NEC Page 72

CHANGE TYPE: Revision (Proposal 4-44a)

CHANGE SUMMARY: Section 230.2 has long required that a building or structure be supplied by only one service. Several exceptions to this rule permit additional services as provided in (A) Special Conditions, (B) Special Occupancies, (C) Capacity Requirements, and (D) Different Characteristics. A new list item, 6, has been added to (A) Special Conditions to permit another service for "enhanced reliability."

2005 CODE: "**230.2 Number of Services.** A building or other structure served shall be supplied by only one service unless permitted in 230.2(A) through (D). For the purpose of 230.40, Exception No. 2 only, underground sets of conductors, 1/0 AWG and larger, running to the same location and connected together at their supply end but not connected together at their load end, shall be considered to be supplying one service.

(A) **Special Conditions.** Additional services shall be permitted to supply the following:

(1) Fire pumps

(2) Emergency systems

(3) Legally required standby systems

(4) Optional standby systems

(5) Parallel power production systems

(6) Systems designed for connection to multiple sources of supply for the purpose of enhanced reliability"

CHANGE SIGNIFICANCE: As with 225.30, this was a proposal by CMP-4 to address "an ongoing issue for a number of Code cycles." The Panel again states, "Redundant supplies are often necessary where supplying critical loads, such as medical laboratories that would not be normally covered under Article 517, banks, or computer loads."

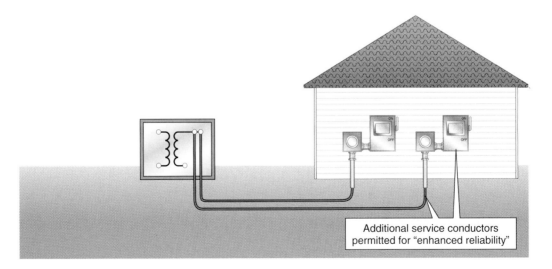

Additional service conductors permitted for "enhanced reliability"

FIGURE 2.27

Similar to 225.30, absent from the section is any guidance on what "systems designed for connection to multiple sources of supply" are. Typically, double-ended switchboards with a tie breaker or switch fit this description. When this equipment is installed at the service position, it often is supplied from different utility transformers in order to achieve additional reliability of supply.

Note that this provision is not intended to relate to emergency systems, legally required standby systems, or optional standby systems as those systems are already provided for in (2), (3), and (4) of this section.

TIE
BREAKER

230.40 Exception No. 1

Number of Service-Entrance Conductor Sets

NEC Page 74

CHANGE TYPE: Revision (Proposal 4-72; Comment 4-44)

CHANGE SUMMARY: Section 230.2 generally requires that a building or structure be supplied by one service. Section 230.40 generally provides that each service drop or service lateral for the "service" supplies only one set of service-entrance conductors. Exception No. 1 has been revised to permit one set of service-entrance conductors to be run to each occupancy or group of occupancies.

2005 CODE: "**230.40 Number of Service-Entrance Conductor Sets.** Each service drop or lateral shall supply only one set of service-entrance conductors.

Exception No. 1: A building ~~with one or more than one occupancy~~ shall be permitted to have one set of service-entrance conductors for each service ~~of different characteristics~~, as defined in 230.2~~(D)~~, run to each occupancy or group of occupancies."

CHANGE SIGNIFICANCE: The substantiation for this proposal supported deleting Exception No. 1 since the previous wording referred to 230.2(D), which already permitted an additional service for "different characteristics." Exception No. 1 then permitted a set of service-entrance conductors for each service *only* if the additional service was installed having "different characteristics."

The Panel rejected the deletion proposal, stating that the exception was necessary: "The exception is needed where the services of different characteristics are required in more than one occupancy."

The exception was revised at the comment stage as shown above. As indicated, the revised exception permits "one set of service-entrance conductors … run to each occupancy or group of occupancies." This

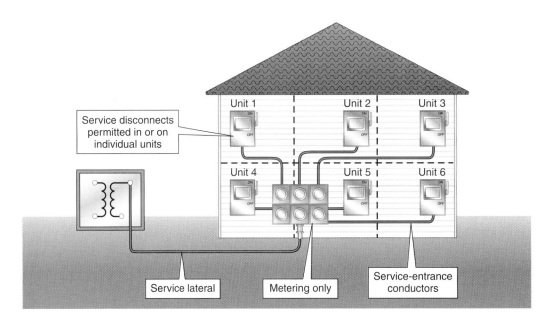

Service disconnects permitted in or on individual units

Service lateral   Metering only   Service-entrance conductors

FIGURE 2.28

revised exception clearly applies to multiple-occupancy buildings. The exception will permit the service drop or service lateral to terminate at the building, perhaps in multimeter equipment, though this is not a requirement to apply the exception. A pull or junction box or wireway could be installed where the transition is made from the service drop or service lateral to service-entrance conductors. The conductors can then be run to each occupancy where the service disconnecting means is located.

The rules in 230.71(A) recognize this concept as it refers to "… each set of service-entrance conductors permitted by 230.40 Exception No 1." Up to six disconnecting means are permitted at the termination of each set of service-entrance conductors. This is likely to be true in strip malls or similar occupancies where a main-lug-only power panelboard is installed at the service equipment. In dwelling units, it is common to install lighting and appliance branch-circuit panelboards that have a main breaker.

Finally, the requirement in 230.70(A)(1) must be complied with: "The service disconnecting means shall be installed at a readily accessible location either outside of a building or structure or inside nearest the point of entrance of the service conductors."

230.71(A)

Maximum Number of Disconnects, General

NEC Page 77

CHANGE TYPE: Revision (Proposal 4-90)

CHANGE SUMMARY: Disconnecting means for Transient Voltage Surge Suppresors installed as a part of listed equipment is not required to be counted as one of the "not more than" six service disconnecting means.

2005 CODE: "**(A) General.** The service disconnecting means for each service permitted by 230.2, or for each set of service-entrance conductors permitted by 230.40, Exception Nos. 1, 3, 4, or 5, shall consist of not more than six switches or sets of circuit breakers, or a combination of not more than six switches and sets of circuit breakers, mounted in a single enclosure, in a group of separate enclosures, or in or on a switchboard. There shall be no more than six sets of disconnects per service grouped in any one location. For the purpose of this section, disconnecting means used solely for power monitoring equipment, <u>transient voltage surge suppressors</u>, or the control circuit of the ground-fault protection system or power-operable service disconnecting means, installed as part of the listed equipment, shall not be considered a service disconnecting means."

CHANGE SIGNIFICANCE: This proposal adds transient voltage surge suppressors (TVSSs) to the list of equipment that is not considered as service disconnecting means. If all the provisions for service discon-

FIGURE 2.29

necting means are executed, up to nine service disconnecting means could be installed without being considered a Code violation if they are "part of the listed equipment." Six disconnecting means could be installed for the normal lighting or power circuits, one for power monitoring equipment, one for transient voltage surge suppressors, and one for the control circuit of the ground-fault protection system or power-operable service disconnecting means.

The Code Panel statement on the comment states, "Permitting a TVSS disconnecting means to not be counted as one of the six switches or circuit breakers does not relax the requirements for TVSS systems to be located on the load side of overcurrent protection. The listing of the control circuit for the GFP system, the TVSS system, or the disconnecting means for power monitoring equipment should incorporate adequate protection for these devices and their related circuits."

Section 285.21(A)(1) requires that TVSS equipment be connected on the load side of a service disconnect overcurrent device unless installed in accordance with 230.82(8). Section 230.82(8) permits TVSS equipment to be connected on the supply side of the service disconnect if it is a part of listed equipment and "suitable overcurrent protection and disconnecting means are provided." In other words, there has to be a service disconnecting means for the TVSS equipment but it is not required to count it as a service disconnecting means.

230.72(B)

Additional Service Disconnecting Means

NEC Page 78

CHANGE TYPE: Revision (Proposal 4-94)

CHANGE SUMMARY: "Emergency systems" has been added to the list of service disconnects required to be located remotely from the normal service disconnects. The intent of this change is to help prevent an incident such as a burn-down of normal service equipment from also destroying the emergency service disconnecting means.

2005 CODE: **"(B) Additional Service Disconnecting Means.** The one or more additional service disconnecting means for fire pumps, emergency systems, ~~for~~ legally required standby, or ~~for~~ optional standby services permitted by 230.2 shall be installed remote from the one to six service disconnecting means for normal service to minimize the possibility of simultaneous interruption of supply."

CHANGE SIGNIFICANCE: This section indicates that the number of services permitted is governed by 230.2. Emergency systems are again included in the list of service disconnects, as they were in the 1996 NEC. An emergency system may be supplied by a separate service per 230.2(A) and doesn't involve taps ahead of the "normal" service to supply emergency systems. A tap ahead of the main is not permitted as a power source for emergency systems per 700.12.

Separate services for emergency systems are permitted by 700.12(D). Separation is required for emergency feeder disconnects by 225.34(B); for fire pump disconnects by 695.4(B); for service drops, laterals, and service conductors for emergency systems by 700.12(D); and for legally required standby systems by 701.11(D).

Separation from all other service disconnecting means is required for emergency system service equipment. However, the service disconnects for an emergency system and a fire pump system are not

FIGURE 2.30

prohibited from being *adjacent* to each other by this section, even though the possibility of interruption of supply by an occurrence in one system may be no less than that provided by a "normal" service disconnecting means.

Section 230.71(A) permits up to six disconnecting means to be installed for each service in 230.2(A). Therefore, 12 disconnecting means could be at the same location if two services are permitted and installed. However, the change to 230.72(B) now requires the disconnecting means for the emergency service to be located remote from the normal service.

While the requirement for separation has been in the Code in various forms for many editions, there is no minimum separation given to satisfy the requirement to "minimize the possibility of simultaneous interruption of supply." If a separate or additional service installed is permitted by 700.12(D) and 230.2(A)(2) for the emergency systems, the service is required to be installed at a different location on the building or structure and may have up to six disconnecting means.

Since no definition of the phrases "remote from" and "to minimize the possibility of simultaneous interruption of supply" exist in the NEC, the authority having jurisdiction for enforcing the Code is responsible for interpreting these requirements and enforcing the terms. Some local authorities having jurisdiction have created local regulations to govern the separation required between normal and emergency services.

230.82(2) and (3)

Equipment Connected to the Supply Side of Service Disconnect
NEC Page 78

CHANGE TYPE: Revision (Proposal 4-106; Comment 4-77)

CHANGE SUMMARY: In 230.82, revisions were made to list item (2) by removing meter disconnect switches and creating a new list item (3) covering only meter disconnect switches. Requirements were also added to (3) for the short-circuit current rating of meter disconnect switches.

2005 CODE: "(2) Meters <u>and</u> meter sockets, ~~or meter disconnect switches~~ nominally rated not in excess of 600 volts, provided all metal housings and service enclosures are grounded.

<u>(3) Meter disconnect switches nominally rated not in excess of 600 volts that have a short-circuit current rating equal to or greater than the available short-circuit current, provided all metal housings and service enclosures are grounded.</u>"

CHANGE SIGNIFICANCE: CMP-4 added the provision to have a "meter disconnect switch" ahead of the service disconnecting means during the 2002 NEC cycle. The need for a disconnect ahead of a watthour meter is driven by the utility desire to disconnect power before a watthour meter is installed or removed.

The individual who submitted this proposal claimed that the permission to add this meter disconnect switch ahead of the meter created a significant problem relative to the short-circuit current rating for the switch. If it was installed without overcurrent protection (which is what the proposer of the 2002 language was advocating), then a short circuit on the service conductors downstream of the switch would likely create a violent failure of the switch. Obviously, this was a big concern for the safety of workers who may be exposed to this potential explosion.

The requirement for the switch to have a short-circuit current rating coordinates with the requirements in 110.9 that the switch must be

FIGURE 2.31

able to safely interrupt the fault current that is available at its line terminals. Since the meter disconnect switch will likely be the first piece of equipment in the circuit, it will be exposed to the greatest short-circuit current supplied by the electric utility. Most electric utilities will furnish the available fault current at the terminals of their transformers. A calculation must then be made to determine the available fault current at the line terminals of the meter disconnect switch. The switch must have a short-circuit current rating suitable for the available fault current at its line terminals.

Since the meter disconnect switch is on the supply side of the service equipment, the authority having jurisdiction (AHJ) will usually require that it be marked "suitable for use as service equipment" (or some acceptable variation of that marking). In addition, the listing requirements for the disconnect switch may require that it be a fusible switch to bear the short-circuit current rating marking. If it is marked "suitable for use as service equipment" and if it is fusible, the AHJ may consider the meter disconnect switch as the service disconnect means. This will require the downstream equipment to comply with the rules for grounding in Article 250. Essentially, all of the equipment downstream from the service disconnecting means is considered to be feeders or branch circuits, and the equipment is not generally permitted to be grounded to the neutral or service grounded conductor. Section 250.142(B) Exception No. 2 permits meter enclosures to be grounded by the grounded circuit conductor on the load side of the service disconnecting means under the conditions specified.

230.82(8)

Equipment Connected to the Supply Side of Service Disconnect

NEC Page 78

CHANGE TYPE: Revision (Proposal 4-108; Comment 4-83)

CHANGE SUMMARY: This revision permits transient voltage surge suppressors (TVSSs) on the line side of service equipment.

2005 CODE: "(8) Ground-fault protection systems <u>or transient voltage surge suppressors</u> where installed as part of listed equipment, if suitable overcurrent protection and disconnecting means are provided."

CHANGE SIGNIFICANCE: This proposal adds TVSSs to the list of equipment that is permitted to be connected on the line side of the service disconnecting means. While the TVSS equipment is now permitted to be connected on either side of the service disconnecting means, this change does not imply that a service disconnecting means and overcurrent protection are not required for TVSS equipment.

Section 285.21(A)(1) requires that TVSS equipment be connected on the load side of a service disconnect overcurrent device unless installed in accordance with 230.82(8). Section 230.82(8) permits TVSS equipment to be connected on the supply side of the service disconnect if it is a part of listed equipment and "suitable overcurrent protection and disconnecting means are provided."

FIGURE 2.32

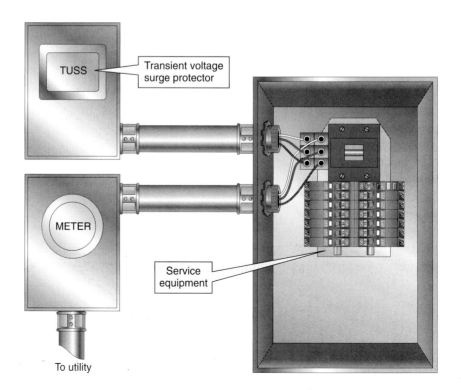

TUSS — Transient voltage surge protector

METER

Service equipment

To utility

240.2

Supervised Industrial Installation
NEC Page 81

CHANGE TYPE: Revision (Proposal 10-9)

CHANGE SUMMARY: The word *feeder* has been added to the definition of *supervised industrial installation*.

2005 CODE: "**Supervised Industrial Installation.** For the purposes of Part VIII, the industrial portions of a facility where all of the following conditions are met:

(1) Conditions of maintenance and engineering supervision ensure that only qualified persons monitor and service the system.

(2) The premises' wiring system has 2500 kVA or greater of load used in industrial process(es), manufacturing activities, or both, as calculated in accordance with Article 220.

(3) The premises has at least one service <u>or feeder</u> that is more than 150 volts to ground and more than 300 volts phase-to-phase."

CHANGE SIGNIFICANCE: The substantiation for this proposal points out that some industrial facilities are supplied by feeders and not necessarily by a service from the utility. Power may only be supplied from on-site or non-utility sources of power, or from another industrial system. In these cases, there will be no service but there will still be one or more feeders.

Part VIII of Article 240 contains several special provisions limited to industrial installations used exclusively for manufacturing or process control activities. Revisions to the rules in other parts of Article 240 include location of overcurrent devices; providing short-circuit, ground fault and overload protection; feeder taps; and special rules for series ratings of overcurrent devices.

FIGURE 2.33

240.20(B) (1), (2), and (3)

Circuit Breaker as Overcurrent Device

NEC Page 84

CHANGE TYPE: Revision (Proposal 10-39)

CHANGE SUMMARY: The term *approved* has been replaced by *identified* in three subdivisions of 240.20(B) to clarify the appropriate use of circuit breaker handle ties.

2005 CODE: "**(B) Circuit Breaker as Overcurrent Device.** Circuit breakers shall open all ungrounded conductors of the circuit both manually and automatically unless otherwise permitted in 240.20(B)(1), (B)(2), and (B)(3).

(1) **Multiwire Branch Circuit.** Except where limited by 210.4(B), individual single-pole circuit breakers, with or without identified approved handle ties, shall be permitted as the protection for each ungrounded conductor of multiwire branch circuits that serve only single-phase line-to-neutral loads.

(2) **Grounded Single-Phase and 3-Wire dc Circuits.** In grounded systems, individual single-pole circuit breakers with identified approved handle ties shall be permitted as the protection for each ungrounded conductor for line-to-line connected loads for single-phase circuits or 3-wire, direct-current circuits.

(3) **3-Phase and 2-Phase Systems.** For line-to-line loads in 4-wire, 3-phase systems or 5-wire, 2-phase systems having a grounded neutral and no conductor operating at a voltage greater than permitted in 210.6, individual single-pole circuit breakers with identified approved handle ties shall be permitted as the protection for each ungrounded conductor."

CHANGE SIGNIFICANCE: The word *approved* is defined as acceptable to the authority having jurisdiction, so an inspector can accept nails, screws, wire, etc as the handle tie.

Identified is defined as suitable for the specific purpose, function, use, environment, application, and so forth, where described in a particular Code requirement.

"Identified" circuit breaker handle ties are readily available from circuit breaker manufacturers and are designed for the purpose expressed above. This revision will assure the use of handle ties that have been designed for this intended function.

FIGURE 2.34

CHANGE TYPE: Revision (Proposal 10-40; Comment 10-40)

CHANGE SUMMARY: A new sentence has been added to state that it is not permitted to round up to the next standard overcurrent device above the ampacity of the conductor when using tap rules.

2005 CODE: "**(B) Feeder Taps.** Conductors shall be permitted to be tapped, without overcurrent protection at the tap, to a feeder as specified in 240.21(B)(1) through (5). <u>The provisions of 240.4(B) shall not be permitted for tap conductors.</u>"

CHANGE SIGNIFICANCE: This proposal was intended to clarify the issue of whether tap conductors, which by definition, do not have overcurrent protection where they originate are required to have an ampacity not less than the rating of the device or overcurrent protection they terminate in.

Ten-foot tap rule conductors are permitted to have overcurrent protection on their supply side of up to ten times the allowable ampacity of the conductor. For example, consider the installation of a 400-ampere feeder, with a tap connected under the 10-foot tap rule. Assuming that all terminations are 75°C, a minimum 8 AWG conductor with an allowable ampacity of 50 amperes is required, as a 10 AWG copper conductor has an allowable ampacity of 35 amperes.

If the tap conductor size is increased to terminate on a panelboard with a 150-ampere rating, a 1/0 AWG conductor would be required (a 1 AWG conductor has an allowable ampacity of 130 amperes), even if the calculated load is 130 amperes or less.

240.21(B)
Feeder Taps
NEC Page 85

FIGURE 2.35

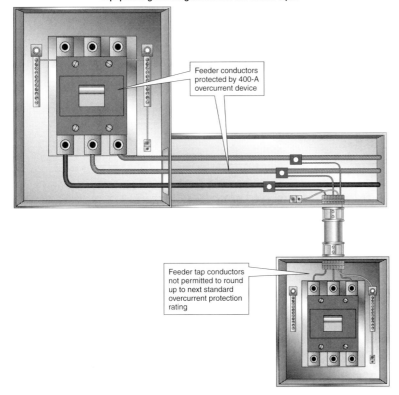

Equipment grounding conductor for feeder tapes

Feeder conductors protected by 400-A overcurrent device

Feeder tap conductors not permitted to round up to next standard overcurrent protection rating

240.21(B) (3)(2)

Taps Supplying a Transformer [Primary Plus Secondary Not Over 7.5 m (25 ft) Long]

NEC Page 85

CHANGE TYPE: Revision (Proposal 10-47)

CHANGE SUMMARY: Editorial changes have been made to the existing text to simplify the verbiage for the tap rule applicable to transformers where a tap is made to supply the primary of a transformer.

2005 CODE: "(2) The conductors supplied by the secondary of the transformer shall have an ampacity that ~~when multiplied by the ratio of the secondary-to-primary voltage, is at least~~ is not less than the value of the primary-to-secondary voltage ratio multiplied by one-third of the rating of the overcurrent device protecting the feeder conductors."

CHANGE SIGNIFICANCE: As seen above, the revised language is simpler and more direct. The change is intended to clarify the requirements for the conductors on the secondary of the transformer without changing them; these rules are important because the primary overcurrent device provides short-circuit and ground-fault protection for the secondary conductors. Overload protection of the secondary conductors is provided by the overcurrent device at the termination of the conductors.

A summary of the rules follows:

(1) The conductors on the primary have an allowable ampacity of at least $1/3$ of the upstream overcurrent device.

(2) The conductors on the secondary have an ampacity not less than $1/3$ of the primary overcurrent protection multiplied by the ratio of the primary-to-secondary ratio of the transformer.

(3) The total length of the primary and secondary conductors is not more than 25 feet.

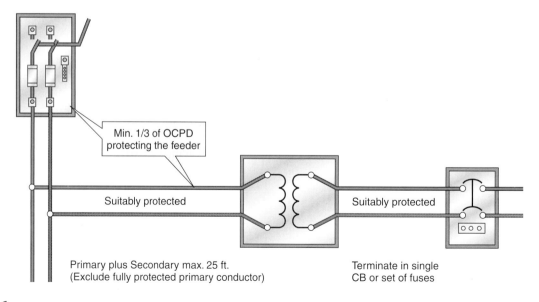

Min. 1/3 of OCPD protecting the feeder

Suitably protected

Suitably protected

Primary plus Secondary max. 25 ft. (Exclude fully protected primary conductor)

Terminate in single CB or set of fuses

FIGURE 2.36

(4) The primary and secondary conductors are protected from physical damage.

(5) The secondary conductors terminate in a single overcurrent device that limits load current to that permitted by 310.15.

As an example, consider a transformer installed to reduce the primary voltage of 480 volts to 208 volts. A 200-ampere overcurrent device is installed protecting the feeder which supplies the tap conductors. What is the minimum ampacity and size of primary and secondary conductors?

Minimum size of primary conductor = 200 amperes ÷ 3 = 66.6 amperes = 4 AWG conductor with 75°C terminations (allowable ampacity of 85 amperes)

Minimum size of secondary conductor = 480 ÷ 208 = 2.3 × 66.6 (200 amperes ÷ 3) = 153 amperes = 2/0 AWG conductor

The maximum overcurrent protection for the conductors on the secondary is 175 amperes, which is the allowable ampacity of the secondary conductors. Keep in mind that this example is used to determine the minimum ampacity of conductors. Overcurrent protection of the transformer is covered in 450.3.

The same editorial change has been made to 240.21(C)(6)(1) for transformer secondary conductors.

240.21(C)

Transformer Secondary Conductors
NEC Page 86

CHANGE TYPE: Revision (Proposal 10-57; Comment 10-49)

CHANGE SUMMARY: Clarification revisions have been made to the first sentence of this section, which clarifies the installation of conductors on the secondary of transformers. A new second sentence forbids rounding up to the next standard size per 240.4(B).

2005 CODE: "**(C) Transformer Secondary Conductors.** Each set of conductors feeding separate loads shall be permitted to be connected to a transformer secondary, without overcurrent protection at the secondary, as specified in 240.21(C)(1) through (6). The provisions of 240.4(B) shall not be permitted for transformer secondary conductors."

CHANGE SIGNIFICANCE: The change to the first sentence is intended to clarify the application of the rule for conductors installed on the secondary of a transformer without overcurrent protection where the conductors originate. The intention of the proposal was that each set of conductors that is installed or connected to the secondary of the transformer be the same size.

The substantiation for this proposal states, "For example, a transformer secondary feeding two unequally-rated overcurrent devices. I believe the NEC intent is to have these two sets of conductors be the same size; however, the existing wording can lead someone to install two sets of differently sized conductors whose combined ampacity is more than the secondary current rating of the transformer."

The new second sentence clarifies that full-size conductors are required on the secondary of the transformer. It is now clear that it is no longer permitted to round up to the next standard rating of overcurrent device when the allowable ampacity of the conductor does not meet or exceed the overcurrent protection on the load end of the

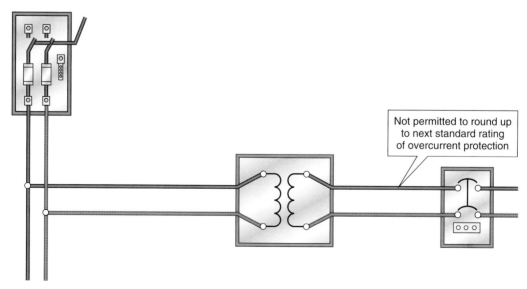

Not permitted to round up to next standard rating of overcurrent protection

FIGURE 2.37

conductor. For example, if a 300-ampere overcurrent device is located at the load end of the conductors, a 350-kcmil copper conductor with an allowable ampacity of 310 amperes must be selected. A 300-kcmil conductor with an allowable ampacity of 285 amperes does not meet or exceed the rating of the overcurrent device and would not be permitted by the new rule.

240.21(C) (2)(1)

Transformer Secondary Conductors Not Over 3 m (10 ft) Long

NEC Page 86

CHANGE TYPE: Revision (Proposal 10-55; Comment 10-50)

CHANGE SUMMARY: Conductors on the secondary of a transformer for the 10-foot maximum length must now be sized similar to the 10-foot tap rule in 240.21(B)(1).

2005 CODE: "**(2) Transformer Secondary Conductors Not Over 3 m (10 ft) Long.** Where the length of secondary conductor does not exceed 3 m (10 ft) and complies with all of the following:

(1) The ampacity of the secondary conductors is:

   a. Not less than the combined <u>calculated</u> ~~computed~~ loads on the circuits supplied by the secondary conductors, and

   b. Not less than the rating of the device supplied by the secondary conductors or not less than the rating of the overcurrent-protective device at the termination of the secondary conductors, <u>and</u>

   c. <u>Not less than one-tenth of the rating of the overcurrent device protecting the primary of the transformer, multiplied by the primary to secondary transformer voltage ratio.</u>

(2) The secondary conductors do not extend beyond the switchboard, panelboard, disconnecting means, or control devices they supply.

(3) The secondary conductors are enclosed in a raceway, which shall extend from the transformer to the enclosure of an enclosed switchboard, panelboard, or control devices or to the back of an open switchboard."

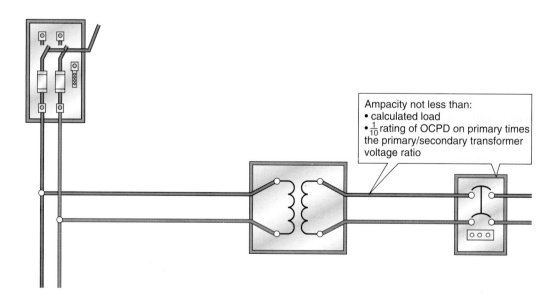

Ampacity not less than:
• calculated load
• $\frac{1}{10}$ rating of OCPD on primary times the primary/secondary transformer voltage ratio

FIGURE 2.38

CHANGE SIGNIFICANCE: This Code Panel action corrects the situation where there previously was no minimum size for conductors up to 10 feet long connected to the secondary of a transformer. As stated in an explanation of the vote on the proposal, "However, the present text of 240.21(C)(2) would allow, for example, a 1500-kVA transformer secondary to be tapped with 14 AWG copper conductors as long as the conductors are not more than ten feet long, protected in a raceway, and terminate in a 15-amp overcurrent device."

For example, consider a 1500-kVA 3-phase transformer installed to reduce the primary voltage of 480 volts to 208 volts. A 2000-ampere overcurrent device is installed on the supply side of the transformer. According to 450.3, overcurrent protection of the secondary is not required, as the overcurrent protection of the primary is less than 125 percent of the transformer full load current: $(1500 \times 1000) \div (480 \times 1.732) = 1804$ FLC $\times 1.25 = 2255$ amperes.

Minimum size of primary conductor = 2000 amperes ÷ 5 sets = 400 amperes = 5, 600-kcmil conductors (rated 420 amperes each) in parallel

Minimum size of secondary conductor = $480 \div 208 = 2.3 \times 200$ (2000 amperes ÷ 10) = 460 amperes = 700-kcmil conductor

Note that this is the minimum conductor size even if a 400-ampere rated panelboard is installed.

240.24

Location in or on Premises, Accessibility

NEC Page 87

CHANGE TYPE: Revision (Proposal 10-67; Comment 10-61)

CHANGE SUMMARY: The rules on the accessibility of overcurrent devices have been updated to include the maximum height permitted for disconnecting means from 404.8(A). Four exceptions to the general requirement apply.

2005 CODE: "**240.24 Location in or on Premises.**
   (A) Accessibility. Overcurrent devices shall be readily accessible and shall be installed so that the center of the grip of the operating handle of the switch or circuit breaker, when in its highest position, is not more than 2.0 m (6 ft 7 in.) above the floor or working platform unless one of the following applies:

   (1) For busways, as described in 368.12.
   (2) For supplementary overcurrent protection, as described in 240.10.
   (3) For overcurrent devices, as described in 225.40 and 230.92.
   (4) For overcurrent devices adjacent to utilization equipment that they supply, access shall be permitted to be by portable means."

CHANGE SIGNIFICANCE: The revision provides clarification as to the maximum mounting height at which overcurrent devices or enclosures that contain overcurrent devices would be considered readily accessible. The proposed revision is consistent with language that already exists in Section 404.8(A) and fills a hole in Article 240, where the maximum height for overcurrent devices that require ready access is not currently specified.

FIGURE 2.39

240.60(D) (New)

Renewable Fuses

NEC Page 89

CHANGE TYPE: New (Proposal 10-78)

CHANGE SUMMARY: This new subdivision permits renewable fuse links only for Class H fuses where there is no evidence of overfusing in existing installations.

2005 CODE: "**(D) Renewable Fuses.** Class H cartridge fuses of the renewable type shall only be permitted to be used for replacement in existing installations where there is no evidence of overfusing or tampering."

CHANGE SIGNIFICANCE: The substantiation for this proposal indicates that renewable-type cartridge fuses have created and continue to create safety issues by overheating when misapplied. The submitter of this proposal noted that all major fuse switch manufacturers have relevant concerns, as noted by the wiring instructions or maintenance material as follows:

- "DO NOT USE RENEWABLE LINK FUSES. The use of renewable link fuses may adversely affect user safety, impair reliability, and will void the warranty."

- "Experience has shown that renewable fuses can cause overheating problems and thus the use of renewable fuses is not recommended."

- "Use of renewable fuses in this switch is not recommended."

- "Renewable link fuses are not recommended."

The concern with renewable type fuses is also recognized by industrial facilities, many of which have put policies in place to prohibit the use of renewable fuses in their electrical systems.

The Code Panel added the reference to Class H to preclude the restriction of future types of current-limiting renewable fuses—such as might be manufactured with polymers or solid state designs—that are in fact renewable.

The Panel stated that it was unaware of the significant safety issues suggested in the substantiation and believes that properly renewed renewable fuses have provided safe, reliable overcurrent protection for decades, when applied within their ratings. The Panel also mentioned the minimal 10,000-ampere interrupting rating associated with Class H renewable fuses.

FIGURE 2.40

240.86

Series Ratings
NEC Page 90

CHANGE TYPE: Revision (Proposal 10-82; Comment 10-72)

CHANGE SUMMARY: This section has been significantly changed to allow series combinations to be determined under engineering supervision for existing installations only. The installation of new equipment with series combination ratings must be tested and marked.

2005 CODE: "**240.86 Series Ratings.** Where a circuit breaker is used on a circuit having an available fault current higher than the ~~its~~ marked interrupting rating by being connected on the load side of an acceptable overcurrent protective device having a higher rating, the circuit breaker shall meet the requirements specified in (A) or (B), and (C) ~~240.86(A) and (B) shall apply~~.

(A) Selected Under Engineering Supervision in Existing Installations. The series rated combination devices shall be selected by a licensed professional engineer engaged primarily in the design or maintenance of electrical installations. The selection shall be documented and stamped by the professional engineer. This documentation shall be available to those authorized to design, install, inspect, maintain, and operate the system. This series combination rating, including identification of the upstream device, shall be field marked on the end use equipment.

~~(A) Marking. The additional series combination interrupting rating shall be marked on the end use equipment, such as switchboards and panelboards.~~

(B) Tested Combinations. The combination of line-side overcurrent device and load-side circuit breaker(s) is tested and marked on the end use equipment, such as switchboards and panelboards.

(C) Motor contribution. [no change from existing (B)]."

CHANGE SIGNIFICANCE: The issue at hand is directly within the scope of the NEC: the practical safeguarding of persons and property. Changes to the proposal at the comment stage clarify that the use of engineering supervision is limited to "existing installations." New installations must use tested and marked equipment. Such testing and marking is performed by the equipment manufacturer.

Since all electrical engineers may not be qualified to make the engineering study and design decisions that are needed, the phrase "The series rated combination devices shall be selected by a licensed professional engineer engaged primarily in the design or maintenance of electrical installations" was added at the comment stage. No engineer in his or her right mind would risk revocation of his or her stamp for the design of an unsafe series rated system. When the interrupting rating of existing installations is subjected to larger values of available short-circuit current than their rating, a very dangerous situation exists. This change is necessary to allow an owner to correct this situation.

The concept of selection of series combination ratings under engineering supervision is directed singularly at existing installations in which the available short-circuit current level has increased to a level that exceeds the interrupting rating of an existing OCPD, creating an extremely dangerous situation for installers, maintainers, and inspectors of the equipment. This increase in available fault current usually is the result of a utility making modifications to its distribution system such as increasing the size of conductors, increasing the capacity of transformers or substations, or supplying the premises from a different substation. The alternative, usually completely replacing distribution equipment, is cost prohibitive and often results in no change of equipment. This change provides a cost-effective method for existing installations to achieve adequate protection for persons and property in the event of a fault that could exceed the interrupting rating of an existing overcurrent device.

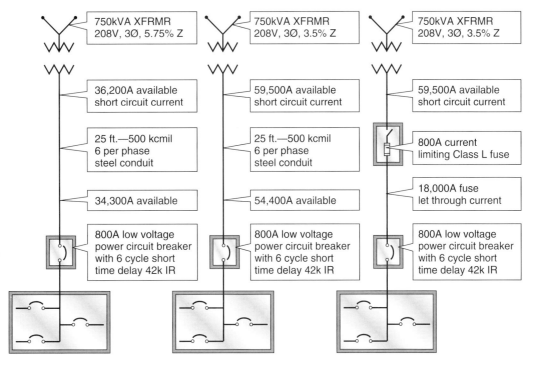

The original installation meets the requirements of 110.9. The transfer is changed by the utility, which increses the available short circuit current. The installation is no longer in compliance with NEC and presents a serious safety hazard. One possible solution is to have a licensed professional engineer select a series rated combination for the existing installation. A current limiting Class L fuse will reduce the available fault curent to 18,000 AMPS. The analysis must be documented and stamped by the professional engineer and the combination rating and identification of the upstream device must be field marked on the equipment to bring the system into compliance with 240.86(A)

FIGURE 2.41

Article 250

Grounding and Bonding
NEC Page 92

CHANGE TYPE: Revision (Proposal 5-37)

CHANGE SUMMARY: The title of Article 250 has been changed from "Grounding" to "Grounding and Bonding."

2005 CODE: "**Grounding and Bonding**"

CHANGE SIGNIFICANCE: The scope of the article clearly covers both grounding and bonding requirements. Adding the term *bonding* to the title of the article is appropriate, and is consistent with the fact that there are generally as many bonding requirements and provisions included within Article 250 as there are grounding requirements, if not more. As seen in Figure 250.4, bonding is directly related to almost every part of Article 250.

FIGURE 2.42

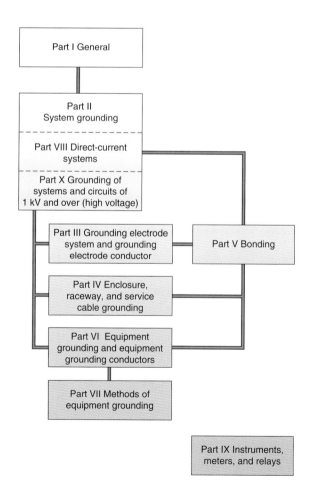

CHANGE TYPE: Revision (Proposal 5-48)

CHANGE SUMMARY: A new phrase has been added to the end of *effective ground-fault current path* to clarify its purpose.

2005 CODE: "**Effective Ground-Fault Current Path.** An intentionally constructed, permanent, low-impedance electrically conductive path designed and intended to carry current under ground-fault conditions from the point of a ground fault on a wiring system to the electrical supply source <u>and that facilitates the operation of the overcurrent protective device or ground-fault detectors on high-impedance grounded systems."</u>

CHANGE SIGNIFICANCE: The identified text was added to the definition to clarify that the purpose of providing an "effective ground-fault current path" is to facilitate the operation of the overcurrent protective device. This is accomplished by providing a low impedance path which will allow sufficient current flow in the event of a fault to facilitate the operation of the overcurrent device. Overcurrent devices are inverse-time. This means that the device operates more quickly if more current is flowing through it. Removing the fault and circuit in the most expeditious manner reduces thermal and magnetic stresses on the equipment.

Similar changes were made to 250.4(A)(5), where the requirement for installing an effective ground-fault current path is provided.

250.2

Effective Ground-Fault Current Path
NEC Page 92

FIGURE 2.43

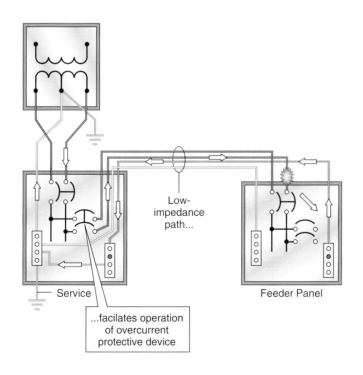

Low-impedance path...

...facilates operation of overcurrent protective device

Service

Feeder Panel

250.8

Connection of Grounding and Bonding Equipment
NEC Page 95

CHANGE TYPE: Revision (Proposal 5-57)

CHANGE SUMMARY: This revision adds "or connection devices" as a type of equipment not permitted to be connected with sheet metal screws.

2005 CODE: "**250.8 Connection of Grounding and Bonding Equipment.** Grounding conductors and bonding jumpers shall be connected by exothermic welding, listed pressure connectors, listed clamps, or other listed means. Connection devices or fittings that depend solely on solder shall not be used. Sheet metal screws shall not be used to connect grounding conductors <u>or connection devices</u> to enclosures."

CHANGE SIGNIFICANCE: This revision should help clarify the requirement for terminals for grounding conductors or bonding jumpers to be attached to enclosures in a suitable method to ensure effective contact and pressure. Previous wording only addressed conductors between a screw and the enclosure, and did not address terminals and lugs that are secured to the enclosure to which grounding or bonding conductors would terminate. This type of connection is in the ground-fault current path and must be effective based on the language in 250.4(A)(5) and 250.4(B)(4).

Sheet metal screws may not provide the required effective path from the conductor to the enclosure. Connection of terminal bars to equipment is generally required to be made with bolts having machine screw threads. Where the screw or bolt threads into an enclosure like a panelboard cabinet, the safety standards require not less than two full threads to be engaged. The thread pitch of sheet metal screws does not provide for connection by two full threads.

FIGURE 2.44

CHANGE TYPE: Revision (Proposal 5-60; Comment 5-42)

CHANGE SUMMARY: The revision to this section requires ground detectors to be installed on systems that can be ungrounded. An exception is provided for systems less than 120 volts to ground.

2005 CODE: "**250.21 Alternating-Current Systems of 50 Volts to 1000 Volts Not Required to Be Grounded.** The following ac systems of 50 volts to 1000 volts shall be permitted to be grounded but shall not be required to be grounded:

(1) Electric systems used exclusively to supply industrial electric furnaces for melting, refining, tempering, and the like

(2) Separately derived systems used exclusively for rectifiers that supply only adjustable-speed industrial drives

(3) Separately derived systems supplied by transformers that have a primary voltage rating less than 1000 volts, provided that all the following conditions are met:

   a. The system is used exclusively for control circuits.

   b. The conditions of maintenance and supervision ensure that only qualified persons service the installation.

   c. Continuity of control power is required.

   d. Ground detectors are installed on the control system.

(4) Other systems that are not required to be grounded in accordance with the requirements of 250.20(B)

250.21

Alternating-Current Systems of 50 Volts to 1000 Volts Not Required to Be Grounded
NEC Page 95

FIGURE 2.45

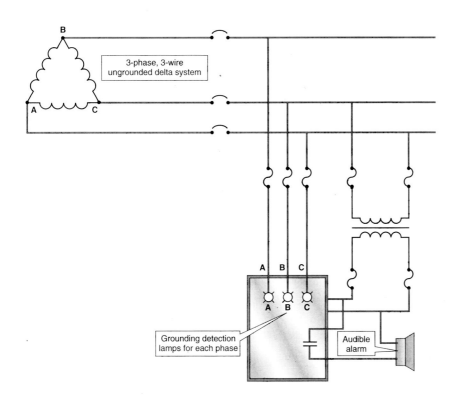

Where an alternating current system is not grounded as permitted in 250.21(1) through (4), ground detectors shall be installed on the system.

Exception: Systems of less than 120 volts to ground as permitted by this Code shall not be required to have ground detectors."

CHANGE SIGNIFICANCE: This revision was one of the more controversial changes made by CMP-5. CMP-5 rejected the original proposal by a vote of 15 to 1. It accepted the proposal at the comment stage by an exact $^2/_3$ vote of 12 to 4.

The substantiation for the proposal states, "This change would be consistent with industry practices when ungrounded systems are installed for continuity of service and minimizing downtime. The Code already requires these detectors for ungrounded control circuits and high-impedance grounded neutral systems. The revision would ensure that there would be a warning of a first phase-to-ground fault on ungrounded systems so it could be corrected prior to the second phase-to-ground fault occurring."

Without a detector, the benefits of the ungrounded system (continuity of service and minimizing outages) are uncertain, and safety for persons and property is compromised by the effects of a second phase-to-ground fault.

CHANGE TYPE: Revision. (Proposal 5-74).

CHANGE SUMMARY: The term *system bonding jumper* has been added to this section because almost all of the characteristics described are identical to those of a main bonding jumper. It is defined in Article 100 and is used throughout 250.30. The text deleted from this section has been moved to become 250.24(B).

2005 CODE: "**250.28 Main Bonding Jumper and System Bonding Jumper.** For a grounded system, main bonding jumpers and system bonding jumpers shall be installed as follows: For a grounded system, an unspliced main bonding jumper shall be used to connect the equipment grounding conductor(s) and the service disconnect enclosure to the grounded conductor of the system within the enclosure for each service disconnect.

Exception No. 1: Where more than one service disconnecting means is located in an assembly listed for use as service equipment, an unspliced main bonding jumper shall bond the grounded conductor(s) to the assembly enclosure.

Exception No. 2: Impedance-grounded-neutral systems shall be permitted to be connected as provided in 250.36 and 250.186.

    (A) Material. Main bonding jumpers and system bonding jumpers shall be of copper or other corrosion-resistant material. A main bonding jumper and a system bonding jumper shall be a wire, bus, screw, or similar suitable conductor.

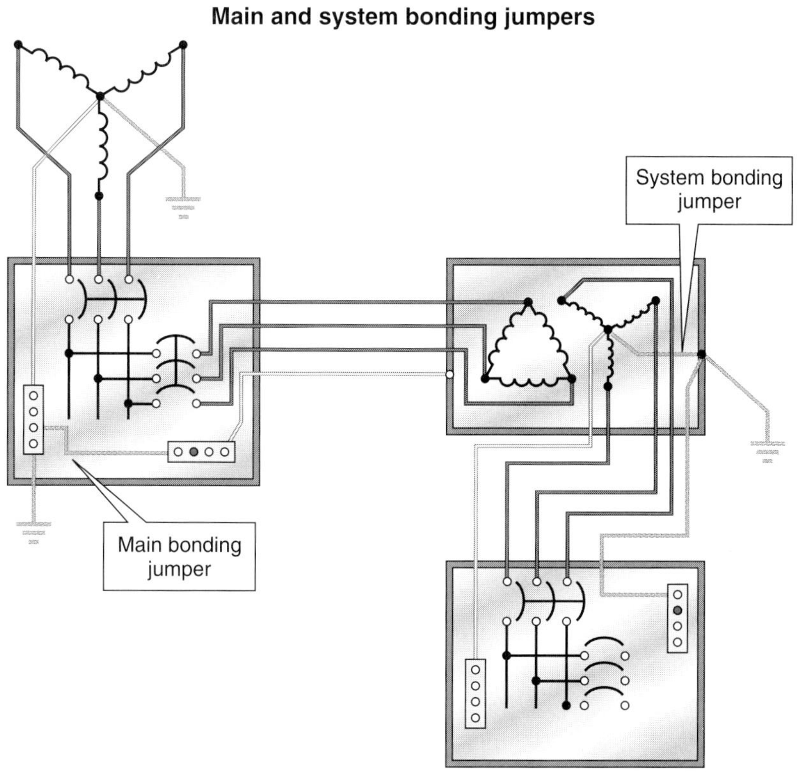

250.28

Main Bonding Jumper and System Bonding Jumper
NEC Page 97

Main and system bonding jumpers

FIGURE 2.46

(B) Construction. Where a main bonding jumper <u>or a system bonding jumper</u> is a screw only, the screw shall be identified with a green finish that shall be visible with the screw installed.

(C) Attachment. Main bonding jumpers <u>and system bonding jumpers</u> shall be attached in the manner specified by the applicable provisions of 250.8.

(D) Size. ~~The main~~ <u>Main</u> bonding ~~jumper~~ <u>jumpers and system bonding jumpers</u> shall not be smaller than the sizes shown in Table 250.66 ~~for grounding electrode conductors~~. Where the ~~service entrance phase~~ <u>supply</u> conductors are larger than 1100 kcmil copper or 1750 kcmil aluminum, the bonding jumper shall have an area that is not less than $12^{1}/_{2}$ percent of the area of the largest phase conductor except that, where the phase conductors and the bonding jumper are of different materials (copper or aluminum), the minimum size of the bonding jumper shall be based on the assumed use of phase conductors of the same material as the bonding jumper and with an ampacity equivalent to that of the installed phase conductors."

CHANGE SIGNIFICANCE: The term *system bonding jumper* is introduced to allow a distinction to be made between the service and separately derived systems. The term *main bonding jumper* is used for service equipment; *system bonding jumper* is used for separately derived systems.

While the Code now makes a distinction between the bonding jumper used at the service and a separately derived system, manufacturers of electrical equipment are not likely to add "system bonding jumper" to their literature and identifying marks. The manufacturer of enclosed panelboards supplies a main bonding jumper with the panelboard. This panelboard is permitted to be used at the service position, as part of a separately derived system and as a building disconnecting means—all with the same bonding jumper installed. The same panelboard is used at the feeder position without the main bonding jumper installed.

A proposal was accepted by CMP-1 to add the following definition of *bonding jumper, system* to Article 100: "The connection between the grounded circuit conductor and the equipment grounding conductor at the separately derived system."

CHANGE TYPE: Revision (Proposal 5-78; Comment 5-52)

CHANGE SUMMARY: The rules on separately derived systems have been reorganized and numbered to improve organization and structure of this section. Several significant changes were incorporated into the reorganization.

2005 CODE: "**250.30 Grounding Separately Derived Alternating-Current Systems.**

(A) Grounded Systems.** A separately derived ac system that is grounded shall comply with 250.30(A)(1) through (A)(8). A grounding connection shall not be made to any grounded circuit conductor on the load side of the point of grounding of the separately derived system except as otherwise permitted in this article."

FPN: See 250.32 for connections at separate buildings or structures, and 250.142 for use of the grounded circuit conductor for grounding equipment.

Exception: Impedance grounded neutral system grounding connections shall be made as specified in 250.36 or 250.186.

(1) **System Bonding Jumper.** An unspliced system bonding jumper in compliance with 250.28(A) through (D) that is sized based on the derived phase conductors shall be used to connect the equipment grounding conductors of the separately derived system to the grounded conductor. This connection shall be made at any single point on the separately derived system from the source to the first system disconnecting means or overcurrent device, or it shall be made at the source of a separately derived system that has no disconnecting means or overcurrent devices.

Exception No. 1: For separately derived systems that are dual fed (double ended) in a common enclosure or grouped together in separate enclosures and employing a secondary tie, a single system

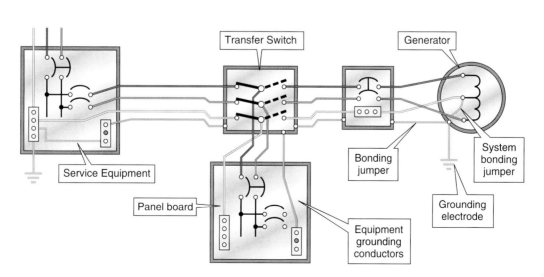

FIGURE 2.47

bonding jumper connection to the tie point of the grounded circuit conductors from each power source shall be permitted.

Exception No. 2: A system bonding jumper at both the source and the first disconnecting means shall be permitted where doing so does not establish a parallel path for the grounded conductor. Where a grounded conductor is used in this manner, it shall not be smaller than the size specified for the system bonding jumper but shall not be required to be larger than the ungrounded conductor(s). For the purposes of this exception, connection through the earth shall not be considered as providing a parallel path.

Exception No. 3: The size of the system bonding jumper for a system that supplies a Class 1, Class 2, or Class 3 circuit, and is derived from a transformer rated not more than 1000 volt-amperes, shall not be smaller than the derived phase conductors and shall not be smaller than 14 AWG copper or 12 AWG aluminum.

(2) Equipment Bonding Jumper Size. Where a bonding jumper of the wire type is run with the derived phase conductors from the source of a separately derived system to the first disconnecting means, it shall be sized in accordance with 250.102(C), based on the size of the derived phase conductors.

(3) Grounding Electrode Conductor, Single Separately Derived System. A grounding electrode conductor for a single separately derived system shall be sized in accordance with 250.66 for the derived phase conductors and shall be used to connect the grounded conductor of the derived system to the grounding electrode as specified in 250.30(A)(7). This connection shall be made at the same point on the separately derived system where the system bonding jumper is installed.

Exception No. 1: Where the system bonding jumper specified in 250.30(A)(1) is a wire or busbar, it shall be permitted to connect the grounding electrode conductor to the equipment grounding terminal bar or bus provided the equipment grounding terminal bar or bus is of sufficient size for the separately derived system.

Exception No. 2: Where a separately derived system originates in listed equipment suitable as service equipment, the grounding electrode conductor from the service or feeder equipment to the grounding electrode shall be permitted as the grounding electrode conductor for the separately derived system provided the grounding electrode conductor is of sufficient size for the separately derived system. Where the equipment ground bus internal to the equipment is not smaller than the required grounding electrode conductor for the separately derived system, the grounding electrode connection for the separately derived system shall be permitted to be made to the bus.

Exception No. 3: A grounding electrode conductor shall not be required for a system that supplies a Class 1, Class 2, or Class 3 circuit and is derived from a transformer rated not more than 1000 volt-amperes, provided the grounded conductor is bonded to the transformer frame or enclosure by a jumper sized in accordance with 250.30(A)(1),

Exception No. 3, and the transformer frame or enclosure is grounded by one of the means specified in 250.134.

(4) Grounding Electrode Conductor, Multiple Separately Derived Systems. Where more than one separately derived system is installed, it shall be permissible to connect a tap from each separately derived system to a common grounding electrode conductor. Each tap conductor shall connect the grounded conductor of the separately derived system to the common grounding electrode conductor. The grounding electrode conductors and taps shall comply with 250.30(A)(4)(a) through (A)(4)(c).

Exception No. 1: Where the system bonding jumper specified in 250.30(A)(1) is a wire or busbar, it shall be permitted to connect the grounding electrode conductor to the equipment grounding terminal bar or bus provided the equipment grounding terminal bar or bus is of sufficient size for the separately derived system.

Exception No. 2: A grounding electrode conductor shall not be required for a system that supplies a Class 1, Class 2, or Class 3 circuit and is derived from a transformer rated not more than 1000 volt-amperes, provided the system grounded conductor is bonded to the transformer frame or enclosure by a jumper sized in accordance with 250.30(A)(1), Exception No. 3, and the transformer frame or enclosure is grounded by one of the means specified in 250.134.

(a) Common Grounding Electrode Conductor Size. The common grounding electrode conductor shall not be smaller than 3/0 AWG copper or 250 kcmil aluminum.

(b) Tap Conductor Size. Each tap conductor shall be sized in accordance with 250.66 based on the derived phase conductors of the separately derived system it serves.

Exception: Where a separately derived system originates in listed equipment suitable as service equipment, the grounding electrode conductor from the service or feeder equipment to the grounding electrode shall be permitted as the grounding electrode conductor for the separately derived system provided the grounding electrode conductor is of sufficient size for the separately derived system. Where the equipment ground bus internal to the equipment is not smaller than the required grounding electrode conductor for the separately derived system, the grounding electrode connection for the separately derived system shall be permitted to be made to the bus.

(c) Connections. All tap connections to the common grounding electrode conductor shall be made at an accessible location by one of the following methods:

(1) A listed connector.

(2) Listed connections to aluminum or copper busbars not less than 6 mm × 50 mm ($^1/_4$ in. × 2 in.). Where aluminum busbars are used, the installation shall comply with 250.64(A).

(3) By the exothermic welding process.

Tap conductors shall be connected to the common grounding electrode conductor in such a manner that the common grounding electrode conductor remains without a splice or joint.

(5) **Installation.** The installation of all grounding electrode conductors shall comply with 250.64(A), (B), (C), and (E).

(6) **Bonding.** Structural steel and metal piping shall be bonded in accordance with 250.104(D).

(7) **Grounding Electrode.** The grounding electrode shall be as near as practicable to and preferably in the same area as the grounding electrode conductor connection to the system. The grounding electrode shall be the nearest one of the following:

(1) Metal water pipe grounding electrode as specified in 250.52(A)(1)

(2) Structural metal grounding electrode as specified in 250.52(A)(2)

Exception No. 1: Any of the other electrodes identified in 250.52(A) shall be used where the electrodes specified by 250.30(A)(7) are not available.

Exception No. 2: Where a separately derived system originates in listed equipment suitable for use as service equipment, the grounding electrode used for the service or feeder equipment shall be permitted as the grounding electrode for the separately derived system.

FPN: See 250.104D for bonding requirements of interior metal water piping in the area served by separately derived systems.

(8) **Grounded Conductor.** Where a grounded conductor is installed and the system bonding jumper is not located at the source of the separately derived system, 250.30(A)(8)(a), (A)(8)(b), and (A)(8)(c) shall apply.

(a) **Routing and Sizing.** This conductor shall be routed with the derived phase conductors and shall not be smaller than the required grounding electrode conductor specified in Table 250.66, but shall not be required to be larger than the largest ungrounded derived phase conductor. In addition, for phase conductors larger than 1100-kcmil copper or 1750-kcmil aluminum, the grounded conductor shall not be smaller than $12^1/_2$ percent of the area of the largest derived phase conductor. The grounded conductor of a 3-phase, 3-wire delta system shall have an ampacity not less than that of the ungrounded conductors.

(b) **Parallel Conductors.** Where the derived phase conductors are installed in parallel, the size of the grounded conductor shall be based on the total circular mil area of the parallel conductors as indicated in this section. Where installed in two or more raceways, the size of the grounded conductor in each raceway shall be based on the size of the ungrounded conductors in the raceway but not smaller than 1/0 AWG.

FPN: See 310.4 for grounded conductors connected in parallel.

(c) **Impedance Grounded System.** The grounded conductor of an impedance grounded neutral system shall be installed in accordance with 250.36 or 250.186.

(B) Ungrounded Systems. The equipment of an ungrounded separately derived system shall be grounded as specified in 250.30(B)(1) and (B)(2).

(1) **Grounding Electrode Conductor.** A grounding electrode conductor, sized in accordance with 250.66 for the derived phase conductors, shall be used to connect the metal enclosures of the derived system to the grounding electrode as specified in 250.30(B)(2). This connection shall be made at any point on the separately derived system from the source to the first system disconnecting means.

(2) **Grounding Electrode.** Except as permitted by 250.34 for portable and vehicle-mounted generators, the grounding electrode shall comply with 250.30(A)(7)."

(Note that the changes are not shown in legislative style with new text underlined and deleted text struck through due to the fact that the changes were too numerous.)

CHANGE SIGNIFICANCE: Many changes were made to this section at the proposal stage and again at the comment stage. Some were made to improve the organization of the requirements, while other changes were substantive. A summary of the changes follows.

- (A) A grounding connection is not permitted on the load side of a separately derived system unless permitted in Article 250. A new FPN is added.

- (A)(1) "System bonding jumper" is used. A new exception deals directly with dual-fed systems rather than pointing to the section on grounding services.

- (A)(2) Reference for bonding jumper size is changed to 250.102(C)

- (A)(3) Rules for a grounding electrode conductor of a single, separately derived system are now located here.

- (A)(4) Rules for multiple, separately derived systems are now located here. The "common" grounding electrode conductor can be no smaller than 3/0. Connections can now include a listed connector, listed connections to an aluminum or copper busbar, and by the exothermic welding process. Irreversible compression connectors are no longer named as they are included in "listed connectors."

- (A)(5) Installation rules refer to 250.64(A), (B), (C), and (E).

- (A)(6) Bonding of structural steel and metal piping must comply with 250.104(D).

- (A)(7) The grounding electrode is required to be the nearest available metal water pipe grounding electrode as specified in 250.52(A)(1) or the structural metal grounding electrode as specified in 250.52(A)(2). Other grounding electrodes are permitted if these are not available.
- (A)(8) Grounded conductor rules include requirements for routing and sizing the conductor(s), installing parallel conductors, and for impedance grounded systems.

(B) Includes requirements for ungrounded separately derived systems.

CHANGE TYPE: Revision (Proposal 5-109; Comments 5-65 and 5-66)

CHANGE SUMMARY: The title of this section has changed from "Two or More Buildings or Structures Supplied from a Common Service" to "Buildings or Structures Supplied by Feeder(s) or Branch Circuit(s)."

2005 CODE: "**250.32 Buildings or Structures Supplied by Feeder(s) or Branch Circuit(s) from a Common Service.**
   (A) Grounding Electrode. Building(s) or structure(s) ~~Where two or more buildings or structures are~~ supplied ~~from a common ac service~~ by ~~a~~ feeder(s) or branch circuit(s) shall have a ~~the~~ grounding electrode~~(s)~~ or grounding electrode system installed in accordance with 250.50. ~~required in Part III of this article at each building or structure~~ The grounding electrode conductor(s) shall be connected in accordance with ~~the manner specified in~~ 250.32(B) or (C). Where there is ~~are~~ no existing grounding electrode~~s~~, the grounding electrode(s) required in 250.50 ~~Part III of this article~~ shall be installed.
   Exception: A grounding electrode ~~at separate buildings or structures~~ shall not be required where only a single ~~one~~ branch circuit supplies the building or structure and the branch circuit includes an equipment grounding conductor for grounding the conductive non-current-carrying parts of all equipment. For the purpose of this section, a multiwire branch circuit shall be considered a single branch circuit.
   (B) Grounded Systems. For a grounded system at the separate building or structure, the connection to the grounding electrode and grounding or bonding of equipment, structures, or frames required to be grounded or bonded shall comply with either 250.32(B)(1) or (B)(2).
   (1) Equipment Grounding Conductor. An equipment grounding conductor as described in 250.118 shall be run with the supply conductors and connected to the building or structure disconnecting means and to the grounding electrode(s). The equipment grounding conductor shall be used for grounding or bonding of equipment, structures, or frames required to be grounded or bonded. The equipment grounding conductor shall be sized

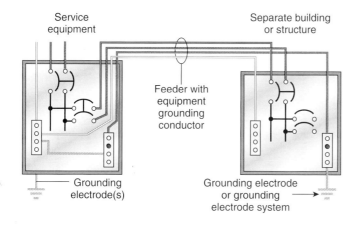

FIGURE 2.48

250.32

Buildings or Structures Supplied by Feeder(s) or Branch Circuit(s)
NEC Page 99

in accordance with 250.122. Any installed grounded conductor shall not be connected to the equipment grounding conductor or to the grounding electrode(s).

(2) Grounded Conductor. Where (1) an equipment grounding conductor is not run with the supply to the building or structure, (2) there are no continuous metallic paths bonded to the grounding system in each building or structure involved, and (3) ground-fault protection of equipment has not been installed on the supply side of the feeder(s) ~~common ac service~~, the grounded circuit conductor run with the supply to the building or structure shall be connected to the building or structure disconnecting means and to the grounding electrode(s) and shall be used for grounding or bonding of equipment, structures, or frames required to be grounded or bonded. The size of the grounded conductor shall not be smaller than the larger of either of the following:

(1) That required by 220.61

(2) That required by 250.122"

CHANGE SIGNIFICANCE: Changes have been made to remove the phrase "from a common service" and replace it with "by feeder(s) or branch circuit(s)." (Some NEC users complained that the phrase "from a common service" was confusing.) Since the service to the premises is only supplied by the electric utility, the buildings supplied from the service are either a feeder or branch circuit. The revisions recognize that buildings or structures may be supplied by feeders from different services and that buildings supplied downstream from the service are supplied by a feeder(s) or branch circuit(s).

CHANGE TYPE: Revision (Proposal 5-115; Comment 5-81)

CHANGE SUMMARY: Revisions to this section require the use of concrete-encased electrodes if they are present. The words "if available" are changed to "are present" to determine when the grounding electrodes have to be bonded together to create the grounding electrode system. A new exception has been added for concrete-encased electrodes in existing buildings.

2005 CODE: "**250.50 Grounding Electrode System.** All grounding electrodes as described in 250.52(A)(1) through (A)(6) that are present at each building or structure served shall be bonded together to form the grounding electrode system. ~~If available on the premises at each building or structure served, each item in 250.52(A)(1) through (A)(6) shall be bonded together to form the grounding electrode system.~~ Where none of these grounding electrodes exist, one or more of the grounding electrodes specified in 250.52(A)(4) through (A)(7) shall be installed and used.

Exception: Concrete-encased electrodes of existing buildings or structures shall not be required to be part of the grounding electrode system where the steel reinforcing bars or rods are not accessible for use without disturbing the concrete."

CHANGE SIGNIFICANCE: The use of the words "if available" in determining when grounding electrodes that are present on the premises at the building or structure supplied are required to be connected to the grounding electrode system have caused confusion among installers and inspectors for many years. A Formal Interpretation was issued

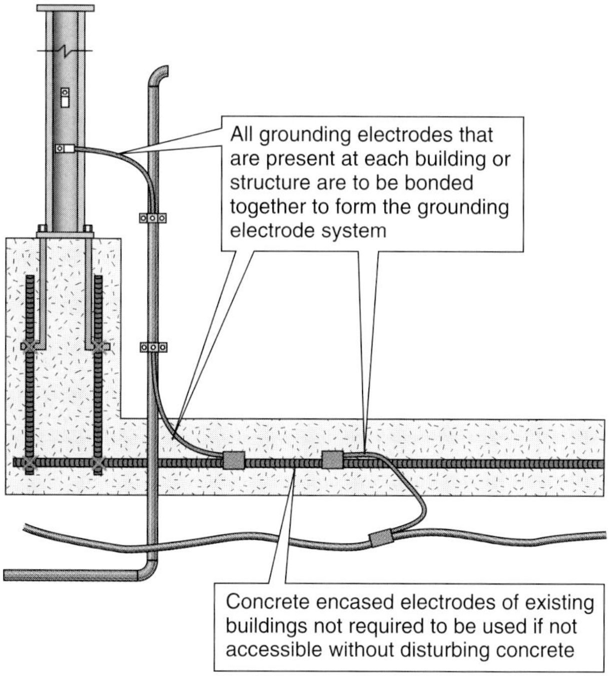

250.50
Grounding Electrode System
NEC Page 101

FIGURE 2.49

All grounding electrodes that are present at each building or structure are to be bonded together to form the grounding electrode system

Concrete encased electrodes of existing buildings not required to be used if not accessible without disturbing concrete

many years ago to indicate that concrete-encased electrodes that were encased when the electrical contractor first arrived at the job site were not considered to be "available." Some NEC users have maintained that this has prevented the use of a proven grounding electrode for too many years.

It has been reported that many jurisdictions have amended this section of the NEC by deleting the words "if available" or otherwise requiring the connection of concrete-encased electrodes. The concrete-encased grounding electrode is a proven effective electrode and is inherent to the construction of most buildings or structures. As a result, it is "present" during the construction process.

By replacing the words "if available" with "if present" the general contractor on construction projects is now required to carefully schedule the installation and bonding of the reinforcing steel, as well as the inspection of the installation, before the concrete is poured.

See 250.52(A)(3) for a description of concrete-encased grounding electrodes.

250.52(A) (2)

Metal Frame of the Building or Structure
NEC Page 101

CHANGE TYPE: Revision (Proposal 5-126; Comment 5-86)

CHANGE SUMMARY: The phrase "where effectively grounded" has been deleted in this section determining when the metal frame of the building is considered a grounding electrode and is thus required to be bonded as a part of the grounding electrode system—"effectively grounded" is subjective in that it is not easily determined. Also, several new methods for creating a grounding electrode consisting of the metal frame of the building or structure are now provided.

2005 CODE: "**(2) Metal Frame of the Building or Structure.** The metal frame of the building or structure, ~~where effectively grounded~~, where any of the following methods are used to make an earth connection:

(1) 3.0 m (10 ft) or more of a single structural metal member in direct contact with the earth or encased in concrete that is in direct contact with the earth

(2) The structural metal frame is bonded to one or more of the grounding electrodes as defined in 250.52(A)(1), (A)(3), or (A)(4)

(3) The structural metal frame is bonded to one or more of the grounding electrodes as defined in 250.52(A)(5) or (A)(6) that comply with 250.56, or

(4) Other approved means of establishing a connection to earth"

CHANGE SIGNIFICANCE: The previous language was considered vague, and as a result was subject to wide and varying interpretations and enforcement. The phrase "effectively grounded" is defined in Article 100; however, no specific measurement factors—such as maximum ohmic resistance value or maximum level of voltage buildup—are provided in the definition. As a result, it was previously somewhat difficult to determine when the metal frame of a building was "effectively

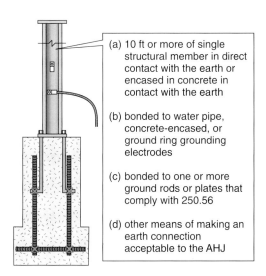

(a) 10 ft or more of single structural member in direct contact with the earth or encased in concrete in contact with the earth

(b) bonded to water pipe, concrete-encased, or ground ring grounding electrodes

(c) bonded to one or more ground rods or plates that comply with 250.56

(d) other means of making an earth connection acceptable to the AHJ

FIGURE 2.50

grounded" and thus was a grounding electrode that was required to be bonded to other grounding electrodes to create the grounding electrode system.

Changes to this section are part of an effort to make the description of the metal frame of the building or structure grounding electrode easy to understand, interpret, and enforce. This revision provides prescriptive language that is easy to interpret, apply, and enforce.

The four methods of determination now provided include:

1. Ten feet or more of a single structural metal member in direct contact with the earth or encased in concrete that is in direct contact with the earth. While this description doesn't mention the rule specifically, it should be understood since the structural metal member is uncoated, or, if coated, is still electrically conductive.

2. The structural metal frame is bonded to one or more grounding electrodes defined in 250.52(A)(1), underground metal water pipe; (A)(3), concrete-encased electrode; or (A)(4), ground ring. This assumes that the other electrodes meet the description in their respective Code section.

3. Bonding to one or more of the grounding electrodes in 250.52(A)(5), rod and pipe electrodes, or (A)(6), plate electrodes. If bonded to these grounding electrodes, the rod and pipe electrodes or plate electrodes are required to meet the 25-ohm maximum resistance rule in 250.56.

4. Other approved means of establishing a connection to earth. This subsection allows the authority having jurisdiction to review and approve the connection of the metal frame of the building or structure to the earth and thus determine when an earth connection has been made.

250.64(C)

Grounding Electrode Conductor Installation, Continuous
NEC Page 103

CHANGE TYPE: Revision (Proposal 5-158; Comment 5-123)

CHANGE SUMMARY: This section has been expanded and reorganized into list format. In addition, this section now permits the use of an aluminum or copper busbar for making connections of the grounding electrode bonding jumpers to the grounding electrode conductor.

2005 CODE: "**(C) Continuous.** ~~The~~ Grounding electrode conductor(s) shall be installed in one continuous length without a splice or joint except as permitted in (1) through (4) ~~unless spliced only by irreversible compression-type connectors listed for the purpose or by the exothermic welding process~~:
~~Exception: Sections of busbars shall be permitted to be connected together to form a grounding electrode conductor.~~

(1) Splicing shall be permitted only by irreversible compression-type connectors listed as grounding and bonding equipment or by the exothermic welding process.

(2) Sections of busbars shall be permitted to be connected together to form a grounding electrode conductor.

(3) Bonding jumper(s) from grounding electrode(s) and grounding electrode conductor(s) shall be permitted to be connected to an

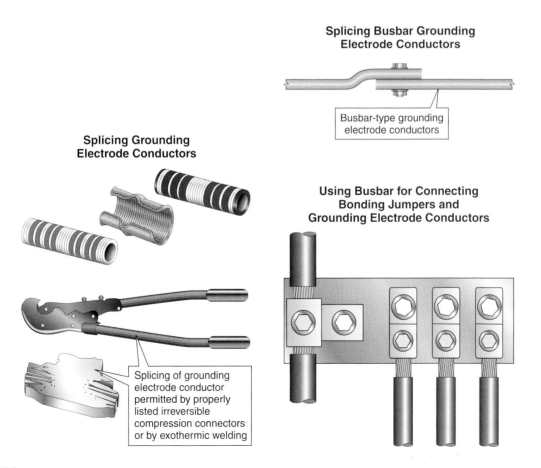

Splicing Busbar Grounding
Electrode Conductors

Busbar-type grounding
electrode conductors

Splicing Grounding
Electrode Conductors

Using Busbar for Connecting
Bonding Jumpers and
Grounding Electrode Conductors

Splicing of grounding
electrode conductor
permitted by properly
listed irreversible
compression connectors
or by exothermic welding

FIGURE 2.51

aluminum or copper busbar not less than 6 mm × 50 mm ($^1/_4$ in. × 2 in.). The busbar shall be securely fastened and shall be installed in an accessible location. Connections shall be made by a listed connector or by the exothermic welding process.

(4) Where aluminum busbars are used, the installation shall comply with 250.64(A)."

CHANGE SIGNIFICANCE: The permission to splice grounding electrode conductors only via an irreversible compression type connector listed as grounding and bonding equipment or by the exothermic welding process remains unchanged. It should be noted that (C)(1) is the only subsection that covers an ordinary "splice" of the grounding electrode conductor.

(C)(2) continues to allow busbars to be connected together to form a grounding electrode conductor. Busbars, of limited lengths and in some cases several lengths, may have to be bolted or welded together to form a grounding electrode conductor.

(C)(3) introduces a new concept for grounding electrode conductors in Section 250.64. Using a busbar to connect grounding electrode conductors was permitted for connecting grounding electrode tap conductors to the common grounding electrode conductor in 250.30 for the 2002 NEC. The new rules for using a busbar for connecting bonding jumpers from grounding electrodes to the grounding electrode conductor include:

1. The busbar must be aluminum or copper not less than $^1/_4$ in. by 2 in.

2. The busbar must be securely fastened in an accessible location.

3. Connections must be made by using a listed connector or the exothermic welding process.

(C)(4), which is also new to this section, mandates that if aluminum busbars are used, the installation is required to comply with the conditions in 250.64(A) for the use of aluminum or copper-clad aluminum conductors.

In several ways, allowing busbars for connections of bonding conductors from grounding electrodes to the grounding electrode conductor is an example of the NEC catching up with industry practice. Electrical engineers have been specifying busbars for making these connections for many years. Authorities having jurisdiction have inspected and approved those installations, so this change should not be a radical departure from previously used methods.

It should be noted that standard listed mechanical connectors are allowed for making conductor connections to the busbar. The connectors are not required to be listed as grounding and bonding equipment in accordance with UL 467. The grounding electrode conductor is not in the effective ground-fault current path, a condition that equipment grounding conductors and some bonding jumpers are tested for.

CHANGE TYPE: Revision (Proposal 5-161; Comment 5-125)

CHANGE SUMMARY: Revisions have been made to this section to incorporate the concept of a common grounding electrode conductor. The rules for sizing and connecting grounding electrode conductor taps have also been updated.

2005 CODE: "**(D) Grounding Electrode Conductor Taps.** Where a service consists of more than a single enclosure as permitted in 230.71(A) ~~230.40, Exception No. 2~~, it shall be permitted to connect taps to the common grounding electrode conductor. Each such tap conductor shall extend to the inside of each such enclosure. The common grounding electrode conductor shall be sized in accordance with 250.66, based on the sum of the circular mil area of the largest ungrounded service-entrance conductors. Where more than one set of service-entrance conductors as permitted by 230.40 Exception No. 2 connect directly to a service drop or lateral, the common grounding electrode conductor shall be sized in accordance with Table 250.66, Note 1. ~~but~~ The tap conductors shall be permitted to be sized in accordance with the grounding electrode conductors specified in 250.66 for the largest conductor serving the respective enclosures. The tap conductors shall be connected to the common grounding electrode conductor in such a manner that the common grounding electrode conductor remains without a splice or joint."

250.64(D)

Grounding Electrode Conductor Taps

NEC Page 104

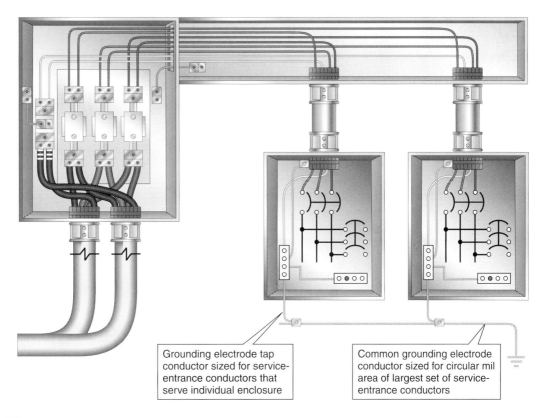

Grounding electrode tap conductor sized for service-entrance conductors that serve individual enclosure

Common grounding electrode conductor sized for circular mil area of largest set of service-entrance conductors

FIGURE 2.52

CHANGE SIGNIFICANCE: The term *common* is added to describe the grounding electrode conductor that is common to the grounding electrode conductor taps and connects to the grounded service conductor for grounded systems or to the service equipment enclosure for ungrounded systems. (The term *common grounding electrode conductor* is not specifically defined.) This change makes this section consistent with 250.30 for separately derived systems.

Common grounding electrode conductors are installed when there are two or more service disconnecting means enclosures and it is desired to have a single grounding electrode conductor extend from the grounding electrode to the area where the service equipment enclosures are located.

The sizing rule for the common grounding electrode conductor is clarified: It is sized differently depending on the method of installing the service-entrance or service-lateral conductors. The common grounding electrode conductor is sized for the service-entrance conductors on the supply side of a wireway where service-entrance conductors connect to the respective service equipment enclosures. See Figure 2.52.

For example, 2500-kcmil service conductors are connected in parallel on the supply side of the wireway. 500 kcmil + 500 kcmil = 1000 kcmil. Refer to Table 250.66 and find a 2/0 copper grounding electrode conductor. This is the minimum size of the common grounding electrode conductor.

The common grounding electrode conductor is sized for the total circular mil area of the service-entrance conductors where the conductors connect to a common service drop. For example, the following sizes of service-entrance conductors are installed to the individual service disconnecting means:

- 3/0, 250 kcmil, 350 kcmil, and 500 kcmil
- 3/0 = 167,800 circular mils from Table 8 of Chapter 9
- 167,800 + 250,000 + 350,000 + 500,000 = 1,267,800 cm

Select a 3/0 AWG common grounding electrode conductor from Table 250.66.

CHANGE TYPE: Revision (Proposals 5-162, 5-164, and 5-190; Comment 5-126)

CHANGE SUMMARY: Revisions have been made to this section to clarify the bonding requirements for ferrous (magnetic) enclosures. Also, bonding requirements from other locations in Article 250 have been incorporated into this section.

2005 CODE: "**(E) Enclosures for Grounding Electrode Conductors.** Ferrous metal enclosures for grounding electrode conductors shall be electrically continuous from the point of attachment to cabinets or equipment to the grounding electrode and shall be securely fastened to the ground clamp or fitting. Nonferrous metal enclosures shall not be required to be electrically continuous. Ferrous metal enclosures that are not physically continuous from cabinets or equipment to the grounding electrode shall be made electrically continuous by bonding each end of the raceway or enclosure to the grounding electrode conductor. Bonding shall apply at each end and to all intervening ferrous raceways, boxes, and enclosures between the service equipment and the grounding electrode. The bonding jumper for a grounding electrode conductor raceway or cable armor shall be the same size as, or larger than, the required enclosed grounding electrode conductor. Where a

250.64(E)

Enclosures for Grounding Electrode Conductors
NEC Page 104

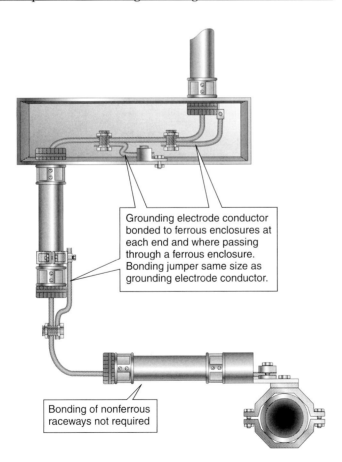

Grounding electrode conductor bonded to ferrous enclosures at each end and where passing through a ferrous enclosure. Bonding jumper same size as grounding electrode conductor.

Bonding of nonferrous raceways not required

FIGURE 2.53

raceway is used as protection for a grounding electrode conductor, the installation shall comply with the requirements of the appropriate raceway article."

CHANGE SIGNIFICANCE: Proposals to revise this section were accepted in principle to clarify that bonding is only required for grounding electrode conductors installed in "ferrous" metal enclosures. (*Ferrous* means "consisting of iron" but is commonly used in place of "magnetic.") Therefore, bonding is not required for nonferrous conduits such as aluminum. Note, however, that there are restrictions on the installation of aluminum conduits, such as not being permitted in corrosive locations and perhaps in direct burial applications. Additional information can be found in the Underwriters Laboratories General Information Directory (White Book) under the heading "Conduit, Rigid Nonferrous Metallic (DYWV)," including "Aluminum conduit used in concrete or in contact with soil requires supplementary corrosion protection."

Also, text has been added to this section from 250.92(A)(3) and 250.102(C) regarding the bonding requirements for metallic enclosures for grounding electrode conductors. Both of these requirements exist in the 2002 NEC and relate directly to the metal enclosures. The word *ferrous* was also added here to make these requirements technically correct.

CHANGE TYPE: New (Proposal 5-176)

CHANGE SUMMARY: This section covers the accessibility of grounding electrode conductor connections. A new exception has been added to exclude exothermic or irreversible compression connections to fire-proofed structural metal from the requirement to be accessible.

2005 CODE: "**(A) Accessibility.** The connection of a grounding electrode conductor or bonding jumper to a grounding electrode shall be accessible.

Exception No. 1: An encased or buried connection to a concrete-encased, driven, or buried grounding electrode shall not be required to be accessible.

<u>*Exception No. 2: An exothermic or irreversible compression connection to fire-proofed structural metal shall not be required to be accessible.*</u>"

CHANGE SIGNIFICANCE: This new exception permits connection of the grounding electrode conductor, by means of exothermic welding or irreversible compression connections, to "fire-proofed structural metal." This connection does not have to be accessible.

Although the structural steel in steel-frame buildings is required to be used as a grounding electrode, and while multiple connections may be required when there are many separately derived systems, the previous edition of the NEC did not recognize the fact that much of this steel is sprayed with fire-proofing that renders the connections inaccessible.

Where a connection is removable, such as when mechanical pressure connectors (lugs) are used, it probably should be made elsewhere or not at all. But if the connection is a permanent connection of a type

250.68(A) Exception No. 2 (New)

Grounding Electrode Conductor and Bonding Jumper Connection to Grounding Electrodes, Accessibility

NEC Page 105

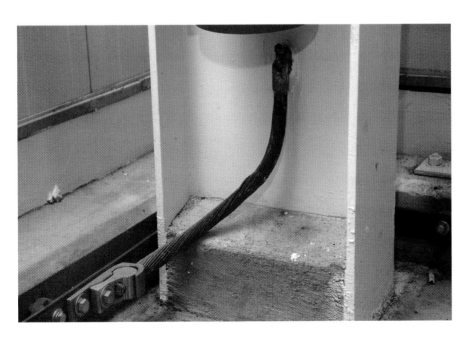

FIGURE 2.54

that could otherwise be buried or encased and connected to ground rings, concrete-encased electrodes, or the like, it should also be permitted when connected to building steel. However, if the steel is required by other standards to be coated or encased in fire-proofing materials, the connection will not be accessible. As it stands, the only way to make these connections accessible is to move them to less substantial steel members that are not fire proofed. This change will allow installers and designers to make the best possible connection to the most substantial steel members.

250.100

Bonding in Hazardous (Classified) Locations
NEC Page 107

CHANGE TYPE: Revision (Proposal 5-189).

CHANGE SUMMARY: Changes have been made to ensure that raceways in hazardous (classified) locations are bonded irrespective of whether an equipment grounding conductor is installed in the raceway.

2005 CODE: "**250.100 Bonding in Hazardous (Classified) Locations.** Regardless of the voltage of the electrical system, the electrical continuity of non-current-carrying metal parts of equipment, raceways, and other enclosures in any hazardous (classified) location as defined in Article 500 shall be ensured by any of the methods specified ~~for services~~ in 250.92(B)(2) through (B)(4) that are approved for the wiring method used. <u>One or more of these bonding methods shall be used whether or not supplementary equipment grounding conductors are installed.</u>"

CHANGE SIGNIFICANCE: This revision was considered after it was reported that some individuals were interpreting the requirement for bonding as being satisfied if an equipment bonding conductor (equipment grounding conductor) was installed with the circuit conductors. However, during a ground-fault condition, current will flow over the raceways and arcing can occur at connection points even with an additional internal equipment grounding conductor installed. The changes indicated above should prevent this from happening.

The term *services* was removed to avoid confusion. Also, the reference to item (1) in 250.92(B) was deleted because bonding to the grounded circuit conductor is not permitted on the load side of the service equipment or the source of a separately derived system.

FIGURE 2.55

250.118 (5)d and (6)e

Types of Equipment Grounding Conductors, Flexible Metal Conduit and Liquidtight Flexible Metal Conduit

NEC Page 110

CHANGE TYPE: Revision (Proposals 5-206 and 5-215; Comments 5-41, 5-190 and 5-192)

CHANGE SUMMARY: The previous requirements of 250.118(5) for flexible metal conduit listed for grounding have been deleted, and the remaining sections have been renumbered. The requirement to add an equipment grounding conductor where flexibility is necessary after installation has been clarified in both 250.118(5)d and 250.118(6)e.

2005 CODE: "**250.118 Types of Equipment Grounding Conductors.**

(5) ~~Flexible metal conduit where both the conduit and fittings are listed for grounding.~~

(5) ~~(6)~~ Listed flexible metal conduit ~~that is not listed for grounding,~~ meeting all the following conditions:

   a. The conduit is terminated in fittings listed for grounding.
   b. The circuit conductors contained in the conduit are protected by overcurrent devices rated at 20 amperes or less.
   c. The combined length of flexible metal conduit and flexible metallic tubing and liquidtight flexible metal conduit in the same ground return path does not exceed 1.8 m (6 ft).
   d. <u>Where used to connect equipment where flexibility is necessary after installation, an equipment grounding conductor shall be installed.</u> ~~The conduit is not installed for flexibility.~~

(6) ~~(7)~~ Listed liquidtight flexible metal conduit meeting all the following conditions:

   a. The conduit is terminated in fittings listed for grounding.
   b. For metric designators 12 through 16 (trade sizes $^3/_8$ through $^1/_2$), the circuit conductors contained in the conduit are protected by overcurrent devices rated at 20 amperes or less.
   c. For metric designators 21 through 35 (trade sizes $^3/_4$ through $1^1/_4$), the circuit conductors contained in the conduit are protected by overcurrent devices rated not more than 60 amperes and there is no flexible metal conduit, flexible metallic tubing, or liquidtight flexible metal conduit in trade sizes metric designators 12 through 16 (trade sizes $^3/_8$ through $^1/_2$) in the grounding path.
   d. The combined length of flexible metal conduit and flexible metallic tubing and liquidtight flexible metal conduit in the same ground return path does not exceed 1.8 m (6 ft).
   e. <u>Where used to connect equipment where flexibility is necessary after installation, an equipment grounding conductor shall be installed.</u> ~~The conduit is not installed for flexibility.~~"

CHANGE SIGNIFICANCE: The previous 250.118(5) was deleted because it referred to a product that was listed by a local testing laboratory and did not comply with the Underwriters Laboratories standard for flexible metal conduit. The product standard and the NEC have long limited the effective length for flexible metal conduit as an equipment grounding conductor to not more than 6 feet, with contained conductors protected not more than 20 amperes.

The previous language regarding when an equipment grounding conductor is required where connected to equipment has been replaced with the phrase, "Where used to connect equipment where flexibility is necessary after installation, an equipment grounding conductor shall be installed."

Flexible metal conduit (FMC) can be installed in two general ways. First, it can be installed as a wiring method through or along framing members and through other spaces such as attics, crawl spaces, and basements. In this type of installation, FMC is secured and supported in accordance with 348.30. General rules mandate that it be securely fastened in place within 12 inches of each box and at intervals not exceeding $4^1/_2$ feet. In this application, flexibility is not required after installation.

Section 348.30 includes four exceptions to the general rule on fastening FMC. Exception No. 2 states, "At terminals where flexibility is required." It seems clear that an equipment grounding conductor would be required through the FMC in this application. Exception No. 3 refers to lengths not exceeding 6 feet from a luminaire terminal connection for tap connections to luminaires. Exception No. 4 permits FMC in lengths not exceeding 6 feet from the last point where the FMC is securely fastened for connections with accessible ceilings to luminaires or other equipment.

The requirement for an equipment grounding conductor installed using liquidtight flexible metal conduit (LFMC) where "flexibility is necessary after installation" is the same as for FMC. See 350.30(A) for the rules on securely fastening LFMC and for exceptions identical to those for FMC.

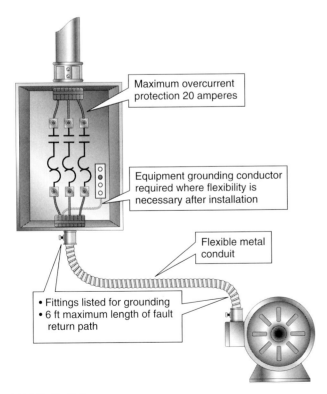
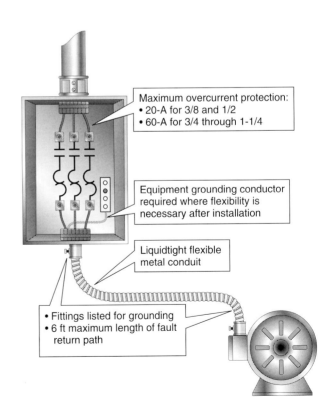

FIGURE 2.56

250.119

Identification of Equipment Grounding Conductors

NEC Page 111

CHANGE TYPE: Revision (Proposal 5-220; Comment 5-196)

CHANGE SUMMARY: This section has been revised to clearly state that conductors with insulation or covering that is green or green with one or more yellow stripes are not permitted to be used as ungrounded (hot) or grounded (neutral) conductors.

2005 CODE: "**250.119 Identification of Equipment Grounding Conductors.** Unless required elsewhere in this *Code*, equipment grounding conductors shall be permitted to be bare, covered, or insulated. Individually covered or insulated equipment grounding conductors shall have a continuous outer finish that is either green or green with one or more yellow stripes except as permitted in this section. Conductors with insulation or individual covering that is green, green with one or more yellow stripes, or otherwise identified as permitted by this section shall not be used for ungrounded or grounded circuit conductors."

CHANGE SIGNIFICANCE: The Code has long required that insulated conductors be green or green with one or more yellow stripes. However, until now, Article 250 has not specifically prohibited the use of conductors so identified for use as ungrounded (hot) or grounded (neutral) conductors.

Electrical inspectors have recently reported that some installers are using conductors with green insulation as ungrounded (hot) conductors by phase-taping the insulation. It has also been reported that the phase tape can come off of the conductor insulation. When this happens, a conductor with green insulation has a potential above ground and presents a shock or flash hazard to electricians. This can be a serious safety issue for workers.

FIGURE 2.57

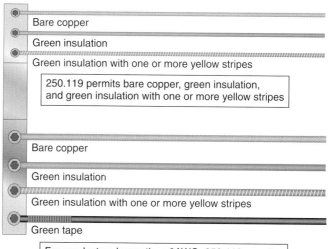

Bare copper

Green insulation

Green insulation with one or more yellow stripes

250.119 permits bare copper, green insulation, and green insulation with one or more yellow stripes

Bare copper

Green insulation

Green insulation with one or more yellow stripes

Green tape

For conductors larger than 6AWG, 250.119 permits bare copper, green insulation, and green insulation with one or more yellow stripes and identification with green marking tape

Since conductor insulations in colors other than green are readily available, there should not be a reason to phase tape green insulation and use it as an ungrounded or grounded conductor.

As revised, this section will permit conductors with green insulation or covering with or without one or more yellow stripes to be used for:

- Equipment grounding conductors
- Equipment bonding jumpers
- Bonding jumpers
- Grounding electrode conductors

250.122 (G)

Size of Equipment Grounding Conductors, Feeder Taps

NEC Page 112

CHANGE TYPE: New (Proposal 5-232)

CHANGE SUMMARY: This addition specifies that the size of the equipment grounding conductor for feeder taps must not be smaller than provided in Table 250.122, based on the rating or setting of the overcurrent device for the feeder. The equipment grounding conductor for the tap is not required to be larger than the tap conductors.

2005 CODE: "**(G) Feeder Taps.** Equipment grounding conductors run with the feeder taps shall not be smaller than shown in Table 250.122 based on the rating of the overcurrent device ahead of the feeder but shall not be required to be larger than the tap conductors."

Equipment grounding conductor size for feeder taps

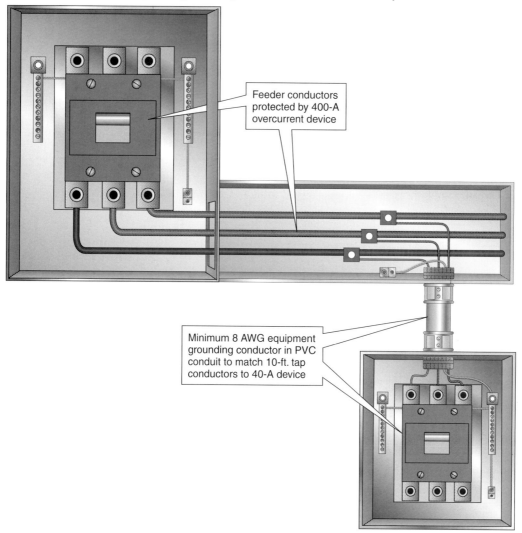

Feeder conductors protected by 400-A overcurrent device

Minimum 8 AWG equipment grounding conductor in PVC conduit to match 10-ft. tap conductors to 40-A device

FIGURE 2.58

CHANGE SIGNIFICANCE: This proposal sought to add clarity to the Code on how to size the equipment grounding conductor when the circuit is a part of a feeder tap. As with other circuits, the equipment grounding conductor is sized based on the rating of the overcurrent device on the supply side. To prevent over-sizing the equipment grounding conductor, this change states that the conductor does not have to be larger than the circuit conductors.

For example, consider a set of conductors installed under the provisions of the 25-foot tap rule in 240.21(B)(2). The tap conductors are required to have an allowable ampacity of not less than $1/3$ the rating of the overcurrent device on the supply side of the feeder. If a 400-ampere overcurrent device protects the feeder tap, the conductors must have an allowable ampacity of $400 \div 3 = 133$ amperes; select 1/0 AWG conductor from Table 310.16. Table 250.122 specifies a 3 AWG copper equipment grounding conductor for a 400-ampere overcurrent device.

For an installation under the provisions of the 10-foot tap rule in 240.21(B)(1), the tap conductors are required to have an allowable ampacity of not less than 10 percent of the rating of the overcurrent device on the supply side of the feeder. If a 400-ampere overcurrent device protects the feeder tap, the conductors must have an allowable ampacity of $400 \div 10 = 40$ amperes; select an 8 AWG conductor from Table 310.16.

Table 250.122 specifies a 3 AWG equipment grounding conductor for a 400-ampere overcurrent device. However the new text also states that the equipment grounding conductor run with feeder taps "shall not be required to be larger than the tap conductors."

250.142 (B) Exception No. 2

Use of Grounded Circuit Conductor for Grounding Equipment

NEC Page 114

CHANGE TYPE: Revision (Proposal 5-237; Comment 5-211)

CHANGE SUMMARY: It is now permitted to ground meter enclosures on the load side of the service disconnecting means to the grounded service conductor if the meter equipment is located "immediately adjacent to" the service equipment. Previously, the word *near* was used.

2005 CODE: *"Exception No. 2: It shall be permissible to ground meter enclosures by connection to the grounded circuit conductor on the load side of the service disconnect if all of the following conditions apply:*

(a) No service ground-fault protection is installed, and

(b) All meter socket enclosures are located <u>immediately adjacent to</u> ~~near~~ the service disconnecting means

(c) The size of the grounded circuit conductor is not smaller than the size specified in Table 250.122 for equipment grounding conductors."

CHANGE SIGNIFICANCE: It is not permitted to ground metal equipment on the load side of the service by means of the grounded conductor. Four exceptions to the rule apply.

In previous editions of the NEC Exception No. 2 has permitted grounding meter enclosures that are "near" the service equipment to the grounded circuit conductor. Since "near" is vague and not definitive, CMP-5 initially voted to delete the exception. The Code Panel reversed this action at the comment stage and voted to replace *near* with the phrase "immediately adjacent to." No maximum distance from the service equipment is provided.

Many plug-in meter sockets are manufactured with the neutral terminal bonded directly to the enclosure. The connection of the grounded service conductor bonds the meter socket on the line side of the service. This type of meter socket is permitted on the load side when it is located "immediately adjacent to" the service disconnecting means.

FIGURE 2.59

250.146 (A)

Connecting Receptacle Grounding Terminal to Box, Surface Mounted Box

NEC Page 115

CHANGE TYPE: Revision (Proposal 5-239; Comment 5-213)

CHANGE SUMMARY: A new sentence requires one of the insulating washers provided on some receptacles to be removed from receptacles to establish direct contact with a metal box.

2005 CODE: "**(A) Surface Mounted Box.** Where the box is mounted on the surface, direct metal-to-metal contact between the device yoke and the box <u>or a contact yoke or device that complies with 250.146(B)</u> shall be permitted to ground the receptacle to the box. <u>At least one of the insulating washers shall be removed from receptacles that do not have a contact yoke or device that complies with 250.146(B) to ensure direct metal-to-metal contact.</u> This provision shall not apply to cover-mounted receptacles unless the box and cover combination are listed as providing satisfactory ground continuity between the box and the receptacle."

CHANGE SIGNIFICANCE: Section 250.146(A) has long permitted the use of direct metal-to-metal contact between a wiring device and a surface metal box to make the bonding connection between the two. Either standard or self-grounding type receptacles have been permitted for this purpose without a bonding jumper from the receptacle to the box.

Section 250.146(B) describes a self-grounding clip or yoke that is designed to ensure an effective grounding circuit between a device yoke and a box. Regardless of the type of box the device is installed on (either flush or surface), this self-grounding feature provides an effective grounding path when there is no direct metal-to-metal contact between the device yoke and the box.

It is not necessary to remove the insulating washers from devices provided with the grounding means described in 250.146(B) because the grounding path is established by the self-grounding clip and the device mounting screws. If a receptacle is provided with insulating washers, it is only necessary to remove one of the washers to ensure an effective grounding path.

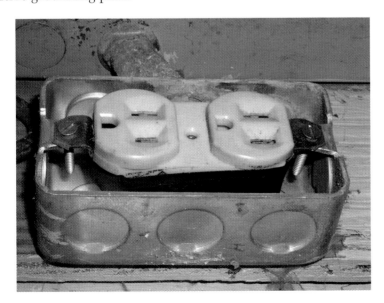

FIGURE 2.60

250.148

Continuity and Attachment of Equipment Grounding Conductors to Boxes

NEC Page 115

CHANGE TYPE: Revision (Proposal 5-248)

CHANGE SUMMARY: Changes to this section include formatting the requirements as subdivisions with titles and editorial improvements to enhance user friendliness of the section.

2005 CODE: **"250.148 Continuity and Attachment of Equipment Grounding Conductors to Boxes.** Where circuit conductors are spliced within a box, or terminated on equipment within or supported by a box, any ~~separate~~ equipment grounding conductor(s) associated with those circuit conductors shall be spliced or joined within the box or to the box with devices suitable for the use <u>in accordance with 250.148(A) through (E).</u>

<u>*Exception: The equipment grounding conductor permitted in 250.146(D) shall not be required to be connected to the other equipment grounding conductors or to the box.*</u>

~~Connections depending solely on solder shall not be used.~~

(A) Connections. <u>Connections and</u> splices shall be made in accordance with 110.14(B) except that insulation shall not be required.

(B) Grounding Continuity. The arrangement of grounding connections shall be such that the disconnection or the removal of a receptacle, luminaire (fixture), or other device fed from the box will not interfere with or interrupt the grounding continuity.

~~*Exception: The equipment grounding conductor permitted in 250.146(D) shall not be required to be connected to the other equipment grounding conductors or to the box.*~~

FIGURE 2.61

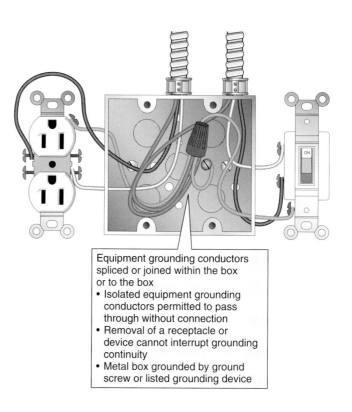

Equipment grounding conductors spliced or joined within the box or to the box
• Isolated equipment grounding conductors permitted to pass through without connection
• Removal of a receptacle or device cannot interrupt grounding continuity
• Metal box grounded by ground screw or listed grounding device

__(C)__ ~~(A)~~ **Metal Boxes.** A connection shall be made between the one or more equipment grounding conductors and a metal box by means of a grounding screw that shall be used for no other purpose or a listed grounding device.

__(D)__ ~~(B)~~ **Non-metallic Boxes.** One or more equipment grounding conductors brought into a non-metallic outlet box shall be arranged so that a connection can be made to any fitting or device in that box requiring grounding.

__(E) Solder.__ Connections depending solely on solder shall not be used."

CHANGE SIGNIFICANCE: This proposal deleted the word *separate* in the opening paragraph because it was unclear what the word referred to. What is a "separate" equipment grounding conductor? One that is not required but is voluntarily installed?

The substantiation for the proposal indicated that it was intended to be an editorial improvement to the section. As previously written, Subsections (A) and (B) seemed to modify and be in conflict with the opening paragraph, which required an equipment grounding connection to a metal box only where circuit conductors are spliced within a box, or terminated on equipment within or supported by a box. The previous wording of (A) required the connection in all arrangements, as did the previous (B).

250.184

Part X Grounding of Systems and Circuits of 1 kV and Over (High Voltage), Solidly Grounded Neutral Systems

NEC Page 117

CHANGE TYPE: Revision (Proposal 5-253; Comment 5-223).

CHANGE SUMMARY: This proposal revised the rules on single-point and multipoint grounding of electrical systems that are 1 kV and over. Multipoint grounding is commonly used by electric utilities. The rules in the NEC mirror, for the most part, the installation practices of most utilities for 1 kV and over systems.

2005 CODE: "**250.184 Solidly Grounded Neutral Systems.** <u>Solidly grounded neutral systems shall be permitted to be either single-point grounded or multigrounded neutral.</u>

(A) Neutral Conductor.

(1) **Insulation Level.** The minimum insulation level for neutral conductors shall be 600 volts.

Exception No. 1: Bare copper conductors shall be permitted to be used for the neutral of service entrances and the neutral of direct buried portions of feeders.

Exception No. 2: Bare conductors shall be permitted for the neutral of overhead portions installed outdoors.

Exception No. 3: The neutral grounded conductor shall be permitted to be a bare conductor if isolated from phase conductors and protected from physical damage.

FPN: See 225.4 for conductor covering within 3.0 m (10 ft) of any building or structure.

(2) **Ampacity.** The neutral conductor shall be of sufficient ampacity for the load imposed on the conductor but not less than $33^1/_3$ percent of the ampacity of the phase conductors.

Exception: In industrial and commercial premises under engineering supervision, it shall be permissible to size the ampacity of the neutral conductor to not less than 20 percent of the ampacity of the phase conductor.

FIGURE 2.62

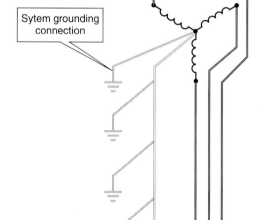

(B) Single-Point Grounded System. Where a single-point grounded neutral system is used, the following shall apply:

(1) A single-point grounded system shall be permitted to be supplied from (a) or (b):

 (a) A separately derived system

 (b) A multigrounded neutral system with an equipment grounding conductor connected to the multigrounded neutral at the source of the single-point grounded system

(2) A grounding electrode shall be provided for the system.

(3) A grounding electrode conductor shall connect the grounding electrode to the system neutral.

(4) A bonding jumper shall connect the equipment grounding conductor to the grounding electrode conductor.

(5) An equipment bonding conductor shall be provided to each building, structure, and equipment enclosure.

(6) A neutral shall only be required where phase to neutral loads are supplied.

(7) The neutral, where provided, shall be insulated and isolated from earth except at one location.

(8) An equipment grounding conductor shall be run with the phase conductors and shall comply with (a), (b), and (c):

 (a) Shall not carry continuous load current

 (b) May be bare or insulated

 (c) Shall have sufficient ampacity for fault current duty

(C) Multigrounded Neutral Systems. Where a multigrounded neutral system is used, the following shall apply:

(1) The neutral of a solidly grounded neutral system shall be permitted to be grounded at more than one point. Grounding shall be permitted at one or more of the following locations:

 (a) Transformers supplying conductors to a building or other structure

 (b) Underground circuits where the neutral is exposed

 (c) Overhead circuits installed outdoors

(2) The multigrounded neutral conductor shall be grounded at each transformer and at other additional locations by connection to a made or existing electrode.

(3) At least one grounding electrode shall be installed and connected to the multigrounded neutral circuit conductor every 400 m (1300 ft).

(4) The maximum distance between any two adjacent electrodes shall not be more than 400 m (1300 ft).

(5) In a multigrounded shielded cable system, the shielding shall be grounded at each cable joint that is exposed to personnel contact."

CHANGE SIGNIFICANCE: Multigrounded neutral systems are standard among utilities and may be necessary for the safety of line personnel. As documented in previous Code proposal substantiations, multigrounding high-voltage systems results in some neutral current flowing through the ground with the potential for serious physiological effects on humans and animals. The Code presently permits this method for premises wiring systems but neither permits nor prohibits the use of single-point grounding systems, which are required for grounded systems below 1 kV.

Industrial plants do not use multigrounded systems because neutral current would flow on water piping, sprinkler systems, process piping, and electrical conduits and would present an undesirable safety hazard to plant personnel.

The purpose of this Code change is to provide a positive alternative for grounding rather than arguing with an inspector that "because it isn't prohibited, it is permitted." No substantive changes have been made to the wording in the 2002 NEC with regard to multigrounded neutral systems.

CHANGE TYPE: Revision (Proposal 5-262; Comment 5-225)

CHANGE SUMMARY: This revision reorganized 280.4 into a list format to improve user friendliness.

2005 CODE: "**280.4 Surge Arrester Selection.**
(A) Circuits of Less Than 1000 Volts. Surge arresters installed on a circuit of less than 1000 volts shall comply with all of the following:

(1) The rating of the surge arrester shall be equal to or greater than the maximum continuous phase-to-ground power frequency voltage available at the point of application.

(2) Surge arresters installed on circuits of less than 1000 volts shall be listed.

(3) Surge arresters shall be marked with a short-circuit current rating and shall not be installed at a point on the system where the available fault current is in excess of that rating.

(4) Surge arresters shall not be installed on ungrounded systems, impedance grounded systems, or corner grounded delta systems unless listed specifically for use on these systems."

CHANGE SIGNIFICANCE: List items (3) and (4) are new material. (3) requires surge arresters under 1000 volts to be marked with a short-circuit current rating. Surge arresters designed for use on circuits of less than 1000 volts are constructed similarly to transient voltage surge suppression (TVSS) products as outlined in Article 285. Surge arresters

280.4(A)
Surge Arrester Selection, Circuits of Less Than 1000 Volts
NEC Page 119

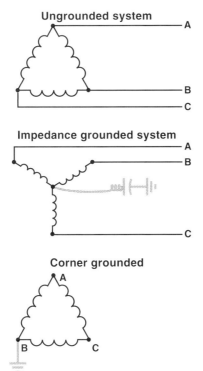

FIGURE 2.63

are required to be tested to higher surge levels than TVSS devices. However, the failure mode of surge arresters under 1000V and TVSS products are similar. Therefore, it is necessary that surge arresters under 1000V have the same short-circuit current rating marked on the arrester to ensure a safe application on the electrical system.

List Item (4) requires that surge arresters not be installed on several systems unless they are specifically listed for use on those systems. The systems specifically listed include:

- Ungrounded systems
- Impedance-grounded systems
- Corner-grounded delta systems

CHANGE TYPE: Revision (Proposal 5-267; Comment 5-230)

CHANGE SUMMARY: This revision mandates that transient voltage surge suppression (TVSS) equipment is not permitted on ungrounded, impedance-grounded, or corner-grounded systems unless specifically listed for such use.

2005 CODE: "**285.3 Uses Not Permitted.** A TVSS shall not be installed ~~used~~ in the following:

(1) Circuits exceeding 600 volts

(2) On ungrounded ~~electrical~~ systems, <u>impedance-grounded systems, or corner-grounded delta systems unless listed specifically for use on these systems</u> ~~as permitted in 250.21~~

(3) Where the rating of the TVSS is less than the maximum continuous phase-to-ground power frequency voltage available at the point of application

> FPN: For further information on TVSSs, see NEMA LS 1-1992, *Standard for Low Voltage Surge Suppression Devices.* The selection of a properly rated TVSS is based on criteria such as maximum continuous operating voltage, the magnitude and duration of overvoltages at the suppressor location as affected by phase-to-ground faults, system grounding techniques, and switching surges."

CHANGE SIGNIFICANCE: The Code Panel received conflicting testimony from experts on both sides of the issue of whether TVSS equipment could be safely used on ungrounded systems, impedance-grounded systems, and corner-grounded systems. While deliberating on the issues involved, the Panel determined that TVSS equipment listed for operation on ungrounded, impedance-grounded, and corner-grounded systems is available or can be made available. As a result, these products can be in compliance with the revised rule only if they are listed for the application, which is the purpose of this revision.

TVSS equipment is nothing more than a very fast switch intended to protect electrical equipment from damaging high-voltage line surges. It does this by clamping overvoltages and routing the transient to ground. A common industry application of TVSS equipment is to protect variable speed drives (VSDs). Lightning arresters are commonly installed at the primary voltage level and are the first line of defense. TVSS equipment is installed at the utilization voltage level.

285.3

Transient Voltage Surge Suppressor, Uses Not Permitted

NEC Page 120

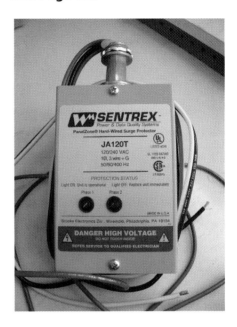

FIGURE 2.64

CHAPTER 3

Wiring Methods and Materials, Articles 300–398

300.4(A) (1) Exception No. 2 (New) and 300.4(A) (2)

Cables and Raceways Through Wood Members, Bored Holes and Notches in Wood

NEC page 123

CHANGE TYPE: New and Revised (Proposals 3-24 and 3-26)

CHANGE SUMMARY: The Panel accepted a new type of nail plate that is less than the standard $1/_{16}$-inch-thick plate that is generally required; this plate must be listed as an equivalent or superior to the $1/_{16}$-inch plate. In addition, the plate must be individually marked so that an inspector can verify its listing even after installation.

2005 CODE: **"(A) Cables and Raceways Through Wood Members.**

(1) **Bored Holes.** In both exposed and concealed locations, where a cable- or raceway-type wiring method is installed through bored holes in joists, rafters, or wood members, holes shall be bored so that the edge of the hole is not less than 32 mm ($1^1/_4$ in.) from the nearest edge of the wood member. Where this distance cannot be maintained, the cable or raceway shall be protected from penetration by screws or nails by a steel plate or bushing, at least 1.6 mm ($1/_{16}$ in.) thick, and of appropriate length and width installed to cover the area of the wiring.

Exception <u>No. 1</u> : Steel plates shall not be required to protect rigid metal conduit, intermediate metal conduit, rigid nonmetallic conduit, or electrical metallic tubing.

<u>*Exception No. 2: A listed and marked steel sleeve less than 1.6 mm ($1/_{16}$ inch) thick that provides equal or better protection against nail and or screw penetration shall be permitted.*</u>

(2) **Notches in Wood.** Where there is no objection because of weakening the building structure, in both exposed and concealed locations, cables or raceways shall be permitted to be laid in notches in wood studs, joists, rafters, or other wood members where the cable or raceway at those points is protected against nails or screws by a steel plate at least 1.6 mm ($1/_{16}$ inch) thick, <u>and of appropriate length and width installed to cover the area of the wiring. The steel plate</u> shall be installed before the building finish is applied."

Exception <u>No. 1</u> (Same as above in 300.4 (A) (1))
Exception <u>No. 2</u> (Same as above in 300.4 (A) (1))

CHANGE SIGNIFICANCE: It was previously maintained that steel plates of $1/_{16}$-inch thickness cause bulging under drywall. Additionally, tests on $1/_{16}$-inch-thick steel have proven that self-tapping screws penetrate this thickness. Therefore, simply specifying a material thickness is not an effective means of preventing damage to cables, wires, or raceways. (A fact finding study was submitted to support the new product.)

The Code Panel chose not to specify a certain thickness of the steel plate nor to specify the minimum hardness of the material. Rather, the Panel is relying on the product safety standard and the listing process to control these issues. The word *marked* was added by the Panel to ensure that an identifier is located on the plate itself, visible after installation.

An identical exception was added to 300.4(A)(2) for protection of non-metallic-sheathed cables and electrical non-metallic tubing through notches in wood; to (B)(2) for non-metallic-sheathed cables and electrical non-metallic tubing through metal framing members; to 300.4(D) for cables and raceways installed parallel to framing members and furring strips; and to 300.4(E) for cables and raceways installed in shallow grooves.

The phrase "and of appropriate length and width installed to cover the area of the wiring" was added to 300.4(A)(2) in order to address how to protect the cables or raceways to which the rule applies when they pass through notches in wood. The concept: The cable or raceway is held close to the surface by being installed in a notch that is not deep enough to allow the cable or raceway to be at least $1^{1}/_{4}$ inches from the surface.

This new requirement also mandates installing protection from penetration by nails or screws of "appropriate length and width" to protect the wiring. No length or width is specified in the rule. As a result, the authority having jurisdiction will determine the "appropriate length and width" of the protection. Since other sections of 300.4 require the protection to be not less than $1^{1}/_{4}$ inches from the stud or other framing member, this will probably also be the minimum width of protection required.

FIGURE 3.1

300.4(D)

Cables and Raceways Parallel to Framing Members <u>and Furring Strips</u>

NEC page 124

CHANGE TYPE: Revision (Proposal 3-17)

CHANGE SUMMARY: The Code Panel added furring strips to this section, clarifying that while furring strips may not be considered framing members, the $1\frac{1}{4}$-inch clearance from any edge subject to nail or screw penetration must be maintained.

2005 CODE: "**(D) Cables and Raceways Parallel to Framing Members <u>and Furring Strips</u>.** In both exposed and concealed locations, where a cable- or raceway-type wiring method is installed parallel to framing members, such as joists, rafters, or studs, <u>or is installed parallel to furring strips</u>, the cable or raceway shall be installed and supported so that the nearest outside surface of the cable or raceway is not less than 32 mm ($1\frac{1}{4}$ in.) from the nearest edge of the framing member <u>or furring strips</u> where nails or screws are likely to penetrate. Where this distance cannot be maintained, the cable or raceway shall be protected from penetration by nails or screws by a steel plate, sleeve, or equivalent at least 1.6 mm ($\frac{1}{16}$ in.) thick."

CHANGE SIGNIFICANCE: Furring strips are usually 1 inch deep by 2 inches wide and are secured to concrete block walls to support sheets of drywall. The drywall is secured to the furring strip with drywall screws or nails.

If a cable- or raceway-type wiring method is installed parallel to the furring strips, the wiring method is just as likely to be damaged from a nail or screw as cables that are installed parallel to other framing members, such as studs or joists. Because of the likelihood of damage, these wiring methods need to have protection identical to other cables routed parallel to framing members. This revision ensures such protection.

FIGURE 3.2

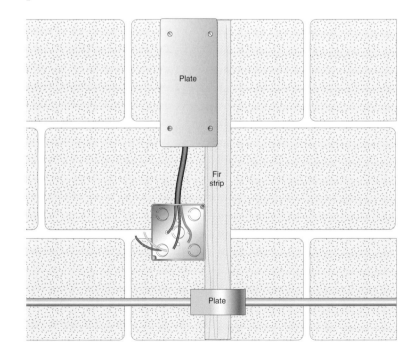

CHANGE TYPE: Relocated (Proposal 3-39a)

CHANGE SUMMARY: The Code Panel moved text related to cables and insulated conductors in underground installations from the former 300.5(D)(5) to a new 300.5(B). The rules should now have broader application.

2005 CODE: "**(B) Listing.** Cables and insulated conductors installed in enclosures or raceways in underground installations shall be listed for use in wet locations."

CHANGE SIGNIFICANCE: This section previously applied only to direct-buried conductors and cables. The requirement for cables and conductors installed in enclosures and raceways in an underground installation to be listed for wet locations should apply to all underground installations. Therefore, this newly located requirement applies to all underground installations.

It should be noted that the definition of *location, wet* in Article 100 indicates that "installations underground or in concrete slabs or masonry in direct contact with the earth" are wet locations. Generally, a conductor with a "W" in its type designation is suitable for wet locations. In addition, 310.8(C)(2) and (3) include the types of conductor insulations suitable for installation in wet locations: "Types MTW, RHW, RHW-2, TW, THW, THW-2, THHW, THHW-2, THWN, THWN-2, XHHW, XHHW-2, ZW; or (3) Of a type listed for use in wet locations."

300.5(B)
Underground Installations, Listing
NEC page 124

FIGURE 3.3

300.6

Protection Against Corrosion and Deterioration

NEC page 126

FIGURE 3.4

CHANGE TYPE: Revision (Proposal 3-51; Comment 3-24)

CHANGE SUMMARY: The Code Panel accepted an addition to the title that clarifies that this section addresses corrosion *and* other deterioration of metallic raceways and other metallic equipment. The section was also rewritten to better cover all possible types of corrosion or deterioration to all metallic systems, as well as to improve the organization of the material.

2005 CODE: "**300.6 Protection Against Corrosion <u>and Deterioration</u>.** ~~Metal~~ Raceways, cable trays, cablebus, auxiliary gutters, cable armor, boxes, cable sheathing, cabinets, elbows, couplings, fittings, supports, and support hardware shall be of materials suitable for the environment in which they are to be installed.

(A) ~~General~~**Ferrous Metal Equipment.** Ferrous <u>metal</u> raceways, cable trays, cablebus, auxiliary gutters, cable armor, boxes, cable sheathing, cabinets, metal elbows, couplings, <u>nipples,</u> fittings, supports, and support hardware shall be suitably protected against corrosion inside and outside (except threads at joints) by a coating of <u>listed</u> ~~approved~~ corrosion-resistant material. ~~such as zinc, cadmium, or enamel. Where protected from corrosion solely by enamel, they shall not be used outdoors or in wet locations as described in 300.6(C). Where boxes or cabinets have an approved system of organic coatings and are marked "Raintight," "Rainproof," or "Outdoor Type," they shall be permitted outdoors.~~ Where corrosion protection is necessary and the conduit is threaded in the field, the threads shall be coated with an approved, electrically conductive, corrosion-resistant compound.

Exception: Stainless steel shall not be required to have protective coatings.

<u>(1) **Protected from Corrosion Solely by Enamel.** Where protected from corrosion solely by enamel, ferrous metal raceways, cable trays, cablebus, auxiliary gutters, cable armor, boxes, cable sheathing, cabinets, metal elbows, couplings, nipples, fittings, supports, and support hardware shall not be used outdoors or in wet locations as described in 300.6(D).</u>

<u>(2) **Organic Coatings on Boxes or Cabinets**. Where boxes or cabinets have an approved system of organic coatings and are marked "Raintight," "Rainproof," or "Outdoor Type," they shall be permitted outdoors.</u>

<u>(3) **In Concrete or in Direct Contact with the Earth.** Ferrous</u> ~~or nonferrous~~ metal raceways, cable armor, boxes, cable sheathing, cabinets, elbows, couplings, nipples, fittings, supports, and support hardware shall be permitted to be installed in concrete or in direct contact with the earth, or in areas subject to severe corrosive influences where made of material approved for the condition, or where provided with corrosion protection approved for the condition.

(B) Non-Ferrous Metal Equipment. Non-ferrous raceways, cable trays, cablebus, auxiliary gutters, cable armor, boxes, cable sheathing, cabinets, elbows, couplings, nipples, fittings, supports, and support hardware embedded or encased in concrete or in direct contact with the earth shall be provided with supplementary corrosion protection.

(C) Nonmetallic Equipment. Nonmetallic raceways, cable trays, cablebus, auxiliary gutters, boxes, cables with a non-metallic outer jacket and internal metal armor or jacket, cable sheathing, cabinets, elbows, couplings, nipples, fittings, supports, and support hardware shall be made of material approved for the condition and shall comply with (C)(1) and (C)(2) as applicable to the specific installation.

(1) Exposed to Sunlight. Where exposed to sunlight, the materials shall be listed as sunlight resistant or shall be identified as sunlight resistant.

(2) Chemical Exposure. Where subject to exposure to chemical solvents, vapors, splashing, or immersion, materials or coatings shall either be inherently resistant to chemicals based on its listing or be identified for the specific chemical reagent.

(D C) Indoor Wet Locations." (No change for Indoor Wet Locations)

CHANGE SIGNIFICANCE: As can be seen above, this section was reorganized to both improve usability and make several substantive changes.

While the addition of "and Deterioration" to the title at first seems substantive, in reality, it is more of an editorial change. *Corrode* means "to wear away gradually usually by chemical action." *Deteriorate* means "to make inferior in quality or value."

Requirements are now clearly separated for ferrous and non-ferrous equipment. Also, the phrase "raceways, cable trays, cablebus, auxiliary gutters, cable armor, boxes, cable sheathing, cabinets, elbows, couplings, nipples, fittings, supports, and support hardware" is intended to be all inclusive and is repeated throughout 300.6.

Clear requirements are now in place for:

- (A) Ferrous metal equipment (means "of iron" and often referred to as "magnetic")
- (A)(1) Ferrous metal equipment protected from corrosion solely by enamel
- (A)(2) Organic coatings on boxes or cabinets
- (A)(3) Ferrous metal equipment in concrete or in direct contact with the earth
- (B) Non-ferrous metal equipment
- (C) Non-metallic equipment
- (C)(1) Non-metallic equipment exposed to sunlight
- (C)(2) Non-metallic equipment exposed to chemicals
- (D) Indoor wet locations

300.11(A) (1)

Securing and Supporting, Fire-Rated Assemblies

NEC page 127

CHANGE TYPE: Revision (Proposal 3-67)

CHANGE SUMMARY: In this section, the Code Panel accepted an added phrase, which clarifies that additional support wires installed for wiring-method support in a fire-rated ceiling may be attached to the ceiling grid.

2005 CODE: "**(1) Fire-Rated Assemblies.** Wiring located within the cavity of a fire-rated floor-ceiling or roof-ceiling assembly shall not be secured to, or supported by, the ceiling assembly, including the ceiling support wires. An independent means of secure support shall be provided <u>and shall be permitted to be attached to the assembly</u>. Where independent support wires are used, they shall be distinguishable by color, tagging, or other effective means from those that are part of the fire-rated design.

Exception: The ceiling support system shall be permitted to support wiring and equipment that have been tested as part of the fire-rated assembly."

CHANGE SIGNIFICANCE: It should be obvious that if you're permitted to install an "independent means of secure support," you are permitted to attach that support to the fire-rated ceiling assembly. This revision clearly permits attachment to the fire rated assembly. In this revision, "independent means of secure support" is required for field-installed wiring that was not submitted, tested, and listed as part of the assembly in order to achieve a fire rating.

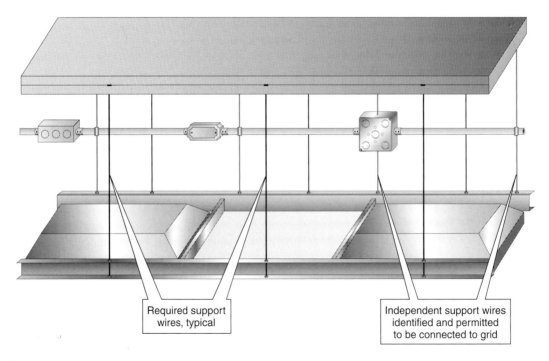

Required support wires, typical

Independent support wires identified and permitted to be connected to grid

FIGURE 3.5

CHANGE TYPE: Revision (Proposal 3-77)

CHANGE SUMMARY: In this revision, handholes have been added to the list of installations not requiring a box for splices, etc.

2005 CODE: "**(L) Manholes <u>and Handhole Enclosures</u>.** Where accessible only to qualified persons, a box or conduit body shall not be required for conductors in manholes <u>or handhole enclosures</u>, except where connecting to electrical equipment. The installation shall comply with the provisions of Part <u>V</u> ~~IV~~ of Article <u>110</u> ~~314~~ for manholes, <u>and 314.30 for handhole enclosures</u>."

CHANGE SIGNIFICANCE: The situation addressed by this proposal is the widespread industry practice of using underground handhole enclosures for the distribution of underground branch wiring in conduit systems, which previously was not permitted by 300.15. The handholes are often used in conjunction with underground PVC conduit as well as other approved wiring methods for the installation of landscape lighting, light poles, and other applications where an above ground box would pose a physical hazard (such as in parks and recreational areas).

Listed wet-location boxes are not designed for immersions during prolonged flooding conditions as experienced in many parts of the country, and have not been the equipment of choice due to the accu-

300.15(L)

Manholes and Handholes
NEC page 129

FIGURE 3.6

mulation of water and potential fault conditions caused when standard splice connections are used within the box.

Listed direct burial splices are required in handholes. The open bottom allows for natural drainage where subject to heavy rain flooding. In industry practice, a conduit is stubbed up within the handhole, a bushing/fitting is installed, the conduit is sealed to prevent foreign entry, continuity is maintained with bonding jumpers, the conductors have insulation suitable for wet locations, splice/taps are made with direct burial listed methods, and covers are secured and grounded if it is metallic.

CHANGE TYPE: Revision (Proposal 3-93)

CHANGE SUMMARY: The Code Panel accepted a deletion of liquidtight flexible metal conduit from this section. The wiring method described will no longer be allowed in manufactured ducts. It was eliminated in "other spaces" used for environmental air in the 2002 NEC.

2005 CODE: "(B) Ducts or Plenums Used for Environmental Air. Only wiring methods consisting of Type MI cable, Type MC cable employing a smooth or corrugated impervious metal sheath without an overall nonmetallic covering, electrical metallic tubing, flexible metallic tubing, intermediate metal conduit, or rigid metal conduit without an overall nonmetallic covering shall be installed in ducts or plenums specifically fabricated to transport environmental air. Flexible metal conduit ~~and liquidtight flexible metal conduit~~ shall be permitted, in lengths not to exceed 1.2 m (4 ft), to connect physically adjustable equipment and devices permitted to be in these ducts and plenum chambers. The connectors used with flexible metal conduit shall effectively close any openings in the connection. Equipment and devices shall be permitted within such ducts or plenum chambers only if necessary for their direct action upon, or sensing of, the contained air. Where equipment or devices are installed and illumination is necessary to facilitate maintenance and repair, enclosed gasketed-type luminaires (fixtures) shall be permitted."

CHANGE SIGNIFICANCE: This revision will complete the action started in the 2002 NEC cycle where all the other wiring methods having a nonmetallic outer jacket were deleted from this section. It was recognized then that there was no limit on the multiple number of 6-foot lengths in "other" ducts or plenums, and there was a similar problem where manufactured ducts are used.

Such ducts or plenums are specifically constructed for transporting environmental air; thus an extra measure of safety is provided to limit in the plenum the amount of combustible material and non-metallic wiring methods capable of creating noxious smoke or fumes.

300.22(B)

Ducts or Plenums Used for Environmental Air
NEC page 131

FIGURE 3.7

310.4

Conductors in Parallel

NEC page 133

CHANGE TYPE: Revision (Proposals 6-6a and 6-10; Comments 6-4 and 6-8)

CHANGE SUMMARY: In this section, "electrically joined at both ends to form a single conductor" was originally deleted, but the phrase "electrically joined at both ends" was added back at the Report on Comments meeting. "Polarity" has also been added throughout to include direct-current applications.

2005 CODE: "**310.4 Conductors in Parallel.** Aluminum, copper-clad aluminum, or copper conductors of size 1/0 AWG and larger, comprising each phase, polarity, neutral, or grounded circuit conductor, shall be permitted to be connected in parallel (electrically joined at both ends to form a single conductor)."

(The four exceptions and FPN are unchanged.)

"The paralleled conductors in each phase, polarity, neutral, or grounded circuit conductor shall comply with all the following:

(1) Be the same length

(2) Have the same conductor material

(3) Be the same size in circular mil area

(4) Have the same insulation type

(5) Be terminated in the same manner

FIGURE 3.8

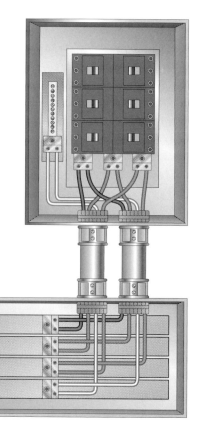

Parallel conductors

The paralleled conductors in each phase, polarity, neutral, or grounded circuit conductor shall:

(1) Be the same length
(2) Have the same conductor material (AL or CU)
(3) Be the same size in circular mil area
(4) Have the same insulation type
(5) Be terminated in the same manner

Where run in separate raceways or cables, the raceways or cables shall have the same physical characteristics

Where run in separate raceways or cables, the raceways or cables shall have the same physical characteristics. Where conductors are in separate raceways or cables, the same number of conductors shall be used in each raceway or cable. Conductors of one phase, polarity, neutral, or grounded circuit conductor shall not be required to have the same physical characteristics as those of another phase, polarity, neutral, or grounded circuit conductor to achieve balance."

CHANGE SIGNIFICANCE: This was a proposal by CMP-6. Their substantiation for the change was quite brief and simple: "The parenthetical phrase does not provide clarity and is not necessary."

While this change may have appeared elementary to the Code Panel, it clears up a huge issue about derating conductors. (The Panel statement on Comment 6-4 refers to possible confusion with the previous language on derating issues.) For example, all the conductors of a 3-phase, 4-wire feeder are installed in a non-metallic wireway. There are three conductors connected together at each end (in parallel) to form a "single conductor." Assuming that the neutral is not considered current-carrying for purposes of derating, is it required to derate in Table 310.15(B)(2)(a) for nine conductors? If the three conductors are truly connected together to form a single conductor, then for the purpose of derating, there are only three current-carrying conductors rather than nine and derating is not required. If it is determined that the neutral conductors *are* current-carrying, four conductors would be counted rather than 12—a big difference in the derating requirements in Table 310.15(B)(2)(a).

Section 310.4 attempts to give guidance as to derating requirements by including the sentence, "Conductors installed in parallel shall comply with the provisions of 310.15(B)(2)(a)." However, if the three conductors per phase are truly being connected together to form a single conductor, derating is not required, since it begins with *four* current-carrying conductors in the same raceway, etc. Most electrical inspectors interpreted the previous Code rules as saying that the individual conductors contribute heat to the installation, and would therefore have determined the number of current-carrying conductors in our example to be nine. Twelve conductors would be used for derating if the neutrals were considered to be current-carrying. (See 310.15(B)(4) for guidance on how to determine if neutral conductors are considered to be current-carrying.) The revision, then, makes it crystal-clear that each conductor of a parallel set is counted individually for derating purposes.

The word *polarity* has been added throughout to clarify that the rules in 310.4 apply to direct-current installations.

The new sentence, "Where conductors are in separate raceways or cables, the same number of conductors shall be used in each raceway or cable," requires a uniformity of installation. It is aimed at ensuring an equal division of the current among all of the conductors connected in parallel. The example cited in Proposal 6-10 is for an 800-ampere installation using three 300-kcmil conductors per phase. Two of the parallel sets of conductors are installed in one conduit and the third is installed in another. Since more than three current-carrying conductors are installed in the first conduit, derating is required here, while it is *not* required for the conductors in the second conduit. This inconsistency results in an uneven distribution of the current among the conductors connected in parallel.

Finally, CMP-6, to make it clear that each conductor counts individually, added the following sentence to 310.15(B)(2)(a); "Each current-carrying conductor of a paralleled set of conductors shall be counted as a current-carrying conductor."

CHANGE TYPE: Deletion (Proposal 6-10a)

CHANGE SUMMARY: This section on the minimum size of conductors has been revised, so all 10 of the previously related exceptions were deleted.

2005 CODE: "**310.5 Minimum Size of Conductors.** The minimum size of conductors shall be as shown in Table 310.5, except as permitted elsewhere in this *Code*."

~~Exception No. 1: For flexible cords as permitted by 400.12.~~

~~Exception No. 2: For fixture wire as permitted by 402.6.~~

~~Exception No. 3: For motors rated 1 hp or less as permitted by 430.22(F).~~

~~Exception No. 4: For cranes and hoists as permitted by 610.14.~~

~~Exception No. 5: For elevator control and signaling circuits as permitted by 620.12.~~

~~Exception No. 6: For Class 1, Class 2, and Class 3 circuits as permitted by 725.27(A) and 725.51, Exception.~~

~~Exception No. 7: For fire alarm circuits as permitted by 760.27(A), 760.51, Exception, and 760.71(B).~~

~~Exception No. 8: For motor-control circuits as permitted by 430.72.~~

~~Exception No. 9: For control and instrumentation circuits as permitted by 727.6.~~

~~Exception No. 10: For electric signs and outline lighting as permitted in 600.31(B) and 600.32(B).~~

CHANGE SIGNIFICANCE: By adding the phrase, "except as permitted elsewhere in this *Code*," the Code Panel determined that the 10 previous exceptions were not needed. This change also recognizes the application of 90.3, which stipulates that "Chapters 1, 2, 3, and 4 apply generally; Chapters 5, 6, and 7 apply to special occupancies, special equipment, or other special conditions. These latter chapters supplement or modify the general rules. Chapters 1 through 4 apply generally except where supplemented or modified by Chapters 5, 6, and 7 for the particular conditions."

310.5

Minimum Size of Conductors
NEC page 134

FIGURE 3.9

Conductor Voltage Rating (Volts)	Minimum Conductor Size (AWG)	
	Copper	Aluminum or Copper-Clad Aluminum
0–2000	14	12
2001–8000	8	8
8001–15,000	2	2
15,000–28,000	1	1
28,000–35,000	1/0	1/0

310.6, Exception

Shielding

NEC page 134

CHANGE TYPE: Revision (Proposal 6-12; Comment 6-15)

CHANGE SUMMARY: In this section's exception, the voltage has been reduced from 8 kV to 2.4 kV for use of unshielded conductors. Cables operated at a voltage higher than 2.4 kV will now be required to be shielded.

2005 CODE: "**310.6 Shielding.** Solid dielectric insulated conductors operated above 2000 volts in permanent installations shall have ozone-resistant insulation and shall be shielded. All metallic insulation shields shall be grounded through an effective grounding path meeting the requirements of 250.4(A)(5) or 250.4(B)(4). Shielding shall be for the purpose of confining the voltage stresses to the insulation.

Exception: Non-shielded insulated conductors listed by a qualified testing laboratory shall be permitted for use up to <u>2400</u> ~~8000~~ volts under the following conditions:

(a) *Conductors shall have insulation resistant to electric discharge and surface tracking, or the insulated conductor(s) shall be covered with a material resistant to ozone, electric discharge, and surface tracking.*

(b) *Where used in wet locations, the insulated conductor(s) shall have an overall non-metallic jacket or a continuous metallic sheath.*

~~(c) Where operated at 5001 to 8000 volts, the insulated conductor(s) shall have a nonmetallic jacket over the insulation. The insulation shall have a specific inductive capacity not greater than 3.6, and the jacket shall have a specific inductive capacity not greater than 10 and not less than 6.~~

(<u>c</u>) *Insulation and jacket thicknesses shall be in accordance with Table 310.63.*"

FIGURE 3.10

CHANGE SIGNIFICANCE: The threshold at which non-shielded conductors are permitted has been reduced from 8 kV to 2.4 kV. As a result of this change, part (c) of the previous exception has been deleted, since it related to conductors operated at from 5 kV to 8 kV.

The substantiation for the proposal alleges that many cable manufacturers specifically recommend against the use of non-shielded cable above 2 kV, but are hesitant to insist on shielding because of concern over having the customer source another supplier. However, several cable manufacturers have experienced arcing problems in customer installations where the cable conductors are separated outside of the outer sheath. These arcing instances are numerous, and present a possible safety hazard.

The previous use of non-shielded 5 kV cables above 2400 volts was reportedly a serious problem, as cables without shielding have a very high failure rate. It has also been mentioned that cables with 90-mil insulation thickness and a jacket are mechanically very weak and subject to mechanical damage. However, if very carefully installed, they will work in a dry location.

310.8(D)

Locations Exposed to Direct Sunlight

NEC page 134

CHANGE TYPE: Revision (Proposal 6-13; Comment 6-23)

CHANGE SUMMARY: This revision lists the insulated conductors and cables that can acceptably be exposed to direct sunlight.

2005 CODE: "**(D) Locations Exposed to Direct Sunlight.** Insulated conductors and cables used where exposed to direct rays of the sun shall <u>comply with one of the following</u>: ~~be of a type listed for sunlight resistance or listed and marked "sunlight resistant."~~

(1) <u>Cables listed, or listed and marked, as being sunlight resistant</u>

(2) <u>Conductors listed, or listed and marked, as being sunlight resistant</u>

(3) <u>Covered with insulating material, such as tape or sleeving, that is listed, or listed and marked, as being sunlight resistant.</u>"

CHANGE SIGNIFICANCE: This section was converted to a list format to improve readability. Covering insulated or covered conductors with UV-resistant material can provide protection from the sun.

Commonly used electrical insulating tape that is sunlight resistant is available.

FIGURE 3.11

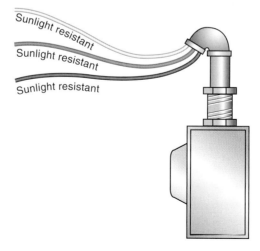

CHANGE TYPE: New (Proposal 6-45)

CHANGE SUMMARY: The Code Panel has added a new Fine Print Note with information on derating for conduits on rooftops that have direct sunlight exposure; 30°F or 17°C is the indicated value to be added to the outdoor ambient temperature to compensate for the direct solar gain.

2005 CODE: "**310.10 Temperature Limitation of Conductors.** No conductor shall be used in such a manner that its operating temperature exceeds that designated for the type of insulated conductor involved. In no case shall conductors be associated together in such a way with respect to type of circuit, the wiring method employed, or the number of conductors that the limiting temperature of any conductor is exceeded.

FPN NO. 1: The temperature rating of a conductor (see Table 310.13 and Table 310.61) is the maximum temperature, at any location along its length, that the conductor can withstand over a prolonged time period without serious degradation. The allowable ampacity tables, the ampacity tables of Article 310 and the ampacity tables of Annex B, the correction factors at the bottom of these tables, and the notes to the tables provide guidance for coordinating conductor sizes, types, allowable ampacities, ampacities, ambient temperatures, and number of associated conductors.

The principal determinants of operating temperature are as follows:

(1) Ambient temperature — ambient temperature may vary along the conductor length as well as from time to time.

(2) Heat generated internally in the conductor as the result of load current flow, including fundamental and harmonic currents.

310.10, FPN No. 2 (New)

Temperature Limitation of Conductors
NEC page 135

FIGURE 3.12

(3) The rate at which generated heat dissipates into the ambient medium. Thermal insulation that covers or surrounds conductors affects the rate of heat dissipation.

(4) Adjacent load-carrying conductors — adjacent conductors have the dual effect of raising the ambient temperature and impeding heat dissipation.

FPN NO. 2: Conductors installed in conduit exposed to direct sunlight in close proximity to rooftops have been shown, under certain conditions, to experience a temperature rise of 17 degrees C (30 degrees F) above ambient temperature on which the ampacity is based."

CHANGE SIGNIFICANCE: The Code Panel accepted a requirement at the ROP meeting that 30°F be added to the ambient temperature where the conduit is exposed to direct sunlight on rooftops. This higher temperature is added to the ambient temperature and can be used to derate the allowable ampacity of conductors from Table 310.16. This proposal was based on a study that included data recordings from a mock-up roof, performed in Las Vegas, Nevada.

After further consideration at the comment meeting, the Panel decided to change the requirement to a non-enforceable Fine Print Note located in 310.10. The Panel states, "While recognizing the study presented to the panel indicates potential for a problem, a more extensive study and evidence needs to be presented before requirements can be added to the Code to address this concern."

CHANGE TYPE: Revised and New (Proposals 6-30 and 6-33)

CHANGE SUMMARY: A new last sentence has been added to 310.15(B)(2)(a) to require that individual conductors of paralleled sets be counted as current-carrying conductors. A new FPN points the Code user to sections that act as exceptions to the rule on derating for more than three current-carrying conductors in a raceway.

2005 CODE: "**(2) Adjustment Factors.**
     (a) More Than Three Current-Carrying Conductors in a Raceway or Cable. Where the number of current-carrying conductors in a raceway or cable exceeds three, or where single conductors or multiconductor cables are stacked or bundled longer than 600 mm (24 in.) without maintaining spacing and are not installed in raceways, the allowable ampacity of each conductor shall be reduced as shown in Table 310.15(B)(2)(a). Each current-carrying conductor of a paralleled set of conductors shall be counted as a current-carrying conductor.

FPN NO. 1: See Annex B, Table B.310.11, for adjustment factors for more than three current-carrying conductors in a raceway or cable with load diversity.

FPN NO. 2: See 366.23(A) for correction factors for conductors in sheet metal auxiliary gutters and 376.22 for correction factors for conductors in metal wireways."

CHANGE SIGNIFICANCE: The addition of the sentence, "Each current-carrying conductor of a paralleled set of conductors shall be counted as a current-carrying conductor," is intended to clarify varying interpretations of whether each conductor of a paralleled set of conductors is required to be counted individually or if all of the conductors count as

310.15(B)(2)(a) and FPN No. 2

Adjustment Factors, More Than Three Current-Carrying Conductors in a Raceway or Cable
NEC page 140

FIGURE 3.13

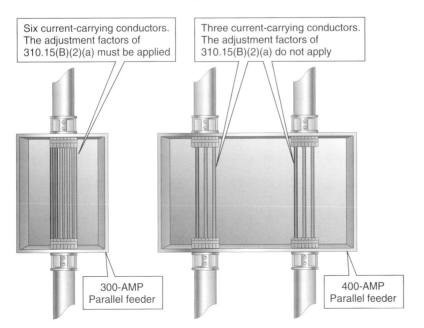

Six current-carrying conductors. The adjustment factors of 310.15(B)(2)(a) must be applied

Three current-carrying conductors. The adjustment factors of 310.15(B)(2)(a) do not apply

300-AMP Parallel feeder

400-AMP Parallel feeder

one for the purpose of derating. Clearly, each conductor of the parallel set is required to be counted individually for the purposes of rating.

One reason for the varying interpretations of the rules was the phrase in 310.4 that previously stated, in part, "shall be permitted to be connected in parallel (electrically joined at both ends to form a single conductor)." Because of this language, some installers maintained that all of the conductors connected in parallel to form a single-phase conductor were counted as one conductor so far as the rating was concerned.

The phrase "to form a single conductor" has been deleted from 310.4. The combination of this deletion and the addition of the sentence on counting conductors to this section should clear up any confusion on how to apply the rules on derating of paralleled conductors.

For example, consider a conduit that contains three sets of paralleled 3/0 conductors for a 208Y/120 3-phase, 4-wire feeder. See the following table for derating for various configurations. for branch circuits in 210.5(C).

Configuration	Conductors if Neural is Counted	Derating Percentage	Conductors if Neural is Not Counted	Derating Percentage
All conductors in single conduit	12	50	9	70
4 conductors in 3 conduits or cables	4	80	3	Not Required
All conductors in metal wireway or auxiliary gutter	12	Not Required	9	Not Required
All conductors in nonmetallic wireway or auxiliary gutter	12	50	9	70

See 301.15(B)(4) for information on determining whether a neutral conductor is required to be counted as current-carrying.

The new FPN will improve the user friendliness of the NEC by pointing out what amounts to exceptions to the general derating rule in 310.15(B)(2)(a) for conductors installed in sheet metal auxiliary gutters or in metal wireways. For metal auxiliary gutters, 366.23(A) indicates that derating is not required so long as the number of current-carrying conductors at any cross section does not exceed 30. A similar statement is included for sheet metal wireways in 376.22. Derating for the number of current-carrying conductors in non-metallic auxiliary gutters and wireways is required, beginning at four current-carrying conductors, because conductors in non-metallic enclosures do not dissipate heat as well as those in sheet-metal enclosures.

CHANGE TYPE: Revision (Proposal 9-5; Comment 9-5)

312.2(A)
Damp and Wet Locations
NEC page 158

CHANGE SUMMARY: This revision adds a requirement that cabinets, cutout boxes, and meter socket enclosures, in wet locations, raceways, or cables that enter above live parts, shall use fittings for wet locations.

2005 CODE: "**(A) Damp and Wet Locations.** In damp or wet locations, surface-type enclosures within the scope of this article shall be placed or equipped so as to prevent moisture or water from entering and accumulating within the cabinet or cutout box, and shall be mounted so there is at least 6 mm ($^1/_4$ in.) airspace between the enclosure and the wall or other supporting surface. Enclosures installed in wet locations shall be weatherproof. <u>For enclosures in wet locations, raceways or cables entering above the level of uninsulated live parts shall use fittings listed for wet locations.</u>

Exception: Nonmetallic enclosures shall be permitted to be installed without the airspace on a concrete, masonry, tile, or similar surface."

CHANGE SIGNIFICANCE: This new sentence brings into the Code requirements that have long been part of the Underwriters Laboratories Product Standards. Installers have been cautioned that for wet locations, any entry must be suitable for the wet location when the entry is above the level of the bussing in the equipment, such as is often the case with panelboards and enclosed switches.

For conduit entries, such as rigid metal conduit and intermediate metal conduit, in a wet location, sealing locknuts are available to seal the entry, as well as sealing-type hubs. Sealing hubs typically have an o-ring that makes contact with the outside of the enclosure to ensure a seal suitable for a wet location.

FIGURE 3.14

Information on the installation of sealing locknuts can be found in the Underwriters Laboratories General Information Directory under Conduit Fittings (DWTT). It reads in part, "Sealing locknuts are intended for use with threaded rigid metal conduit and intermediate metal conduit with one sealing locknut on the outside or inside and either an ordinary locknut or sealing locknut on the opposite side of the enclosure for wet locations or liquid-tight applications. Sealing locknuts may also be used with listed wet location or liquid-tight fittings where so marked on the fitting carton."

Other suitable fittings for entry into enclosures in wet locations include the hub furnished by the manufacturer or those available as an accessory of the enclosure.

CHANGE TYPE: New (Proposal 9-14; Comment 9-7)

CHANGE SUMMARY: This proposal adds requirements for repairing surfaces adjacent to flush-type panelboards that are identical to the rules in 314.21 for boxes.

2005 CODE: "**312.4 Repairing Plaster and Drywall or Plasterboard.** Plaster, drywall, or plasterboard surfaces that are broken or incomplete shall be repaired so there will be no gaps or open spaces greater than 3 mm ($^1/_8$ in.) at the cabinet or cutout box employing a flush-type cover."

CHANGE SIGNIFICANCE: The addition of this new section copies the rules on repairing openings around outlet boxes from 314.21 to Article 312. The new rule applies to flush-type cabinets commonly used for panelboards. The same maximum $^1/_8$-in. space applies.

Absent from the rule is who is responsible for repairing openings with excessive space between the enclosure and the building wall surface. This usually depends on what trades are on the job site as well as the views of the owner or construction manager. If plasterers or sheet rock finishers are present, they will usually be asked to make the repairs. The electrical inspector who enforces the rule will usually write a correction notice to the purchaser of the permit rather than attempt to find out who will be making the repairs.

312.4 (New)

Repairing Plaster and Drywall or Plasterboard
NEC page 158

FIGURE 3.15

Article 314

Outlet, Device, Pull, and Junction Boxes; Conduit Bodies; Fittings and Handhole Enclosures
NEC page 161

CHANGE TYPE: Revision (Proposals 9-15, 9-16, 9-17, and 9-18)

CHANGE SUMMARY: The rules on manholes have been moved from Article 314 to become Part V of Article 110. "Handhole enclosures" has been added to the title of Article 314 and its scope.

2005 CODE: "**314.1 Scope.** This article covers the installation and use of all boxes and conduit bodies used as outlet, device, junction, or pull boxes, depending on their use, and handhole enclosures ~~manholes and other electric enclosures intended for personnel entry~~. Cast, sheet metal, non-metallic, and other boxes such as FS, FD, and larger boxes are not classified as conduit bodies. This article also includes installation requirements for fittings used to join raceways and to connect raceways and cables to boxes and conduit bodies."

CHANGE SIGNIFICANCE: The action to move Part IV on manholes from Article 314 to become Part V of Article 110 is a result of a decision by the Technical Correlating Committee on the organization of the NEC. The substantiation offered includes:

- The working clearance and safety requirements of electrical manholes and related fire resistivity are appropriately located with other relative information in Article 110.

- The provisions of current Part IV in Article 314 were conditional, just like the requirements in Article 110; e.g., the requirements are only applicable where the space "is likely to require examination, adjustment, servicing, or maintenance while energized."

- 314.52 addressed "cabling work space."

FIGURE 3.16

- 314.53 addressed "equipment work space."
- 314.54 only addressed bending space for conductors through requirements in Article 314, Part V, and did make specific requirements in Part III.
- 314.55 addressed access to manholes.
- 314.56 addressed access to vaults and tunnels.
- 314.58 addressed guarding by referencing Article 110.

314.16(B) (1)

Box Fill Calculations, Conductor Fill

NEC page 162

CHANGE TYPE: Revision (Proposal 9-27; Comment 9-65)

CHANGE SUMMARY: A requirement has been added to count certain unbroken conductors for box volume purposes.

2005 CODE: "**(B) Box Fill Calculations.** The volumes in paragraphs 314.16(B)(1) through (B)(5), as applicable, shall be added together. No allowance shall be required for small fittings such as locknuts and bushings.

(1) Conductor Fill. Each conductor that originates outside the box and terminates or is spliced within the box shall be counted once, and each conductor that passes through the box without splice or termination shall be counted once. <u>A looped, unbroken conductor not less than twice the minimum length required for free conductors and 300.14 shall be counted twice.</u> The conductor fill shall be computed using Table 314.16(B). A conductor, no part of which leaves the box, shall not be counted.

Exception: An equipment grounding conductor or conductors or not over four fixture wires smaller than 14 AWG, or both, shall be permitted to be omitted from the calculations where they enter a box from a domed luminaire (fixture) or similar canopy and terminate within that box."

CHANGE SIGNIFICANCE: Loops, coils, and interconnecting wire of a device in a box, that does not leave the box, occupy space. A box can be jammed up with the 2002 Code provisions. This can create a serious fire hazard when an electronic device is installed in a box without adequate room for its heat dissipation. The usage of electronic devices is

FIGURE 3.17

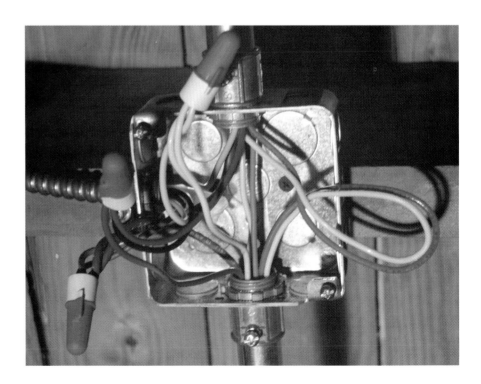

a common practice and is no longer an isolated event; the free space issue had to be addressed by the 2005 Code.

This revision properly distinguishes between a small loop left to assist wire pulling and dressing, and a large loop left to allow cutting in the middle and then adding a splice or a device.

The minimum length of free conductor provided in 300.14 for connection to a device is 6 inches. In addition, where the opening is less than 8 inches in any dimension, the conductor must extend not less than 3 inches from the front. So, for application of the new rule in 314.16(B)(1), if a loop of conductor is not less than twice the length required in 300.14, the loop must be counted as two conductors. This will ensure adequate space in the enclosure. Conductors that are shorter are counted as one conductor.

314.23(B)(1)

Nails and Screws
NEC page 164

CHANGE TYPE: Revision (Proposal 9-50)

CHANGE SUMMARY: In order to avoid damage to conductor insulation, screw threads are no longer permitted to be exposed in a box unless adequately protected.

2005 CODE: "(1) Nails and Screws. Nails and screws, where used as a fastening means, shall be attached by using brackets on the outside of the enclosure, or they shall pass through the interior within 6 mm ($^1/_4$ in.) of the back or ends of the enclosure. Screws shall not be permitted to pass through the box unless exposed threads in the box are protected using approved means to avoid abrasion of conductor insulation."

CHANGE SIGNIFICANCE: This proposal originated with the National Electrical Manufacturers Association, which maintained that coarse screw threads found on most screws used for mounting boxes, when left exposed inside a box, present a severe abrasion hazard to conductor insulation.

FIGURE 3.18

CHANGE TYPE: Relocation (9-57)

CHANGE SUMMARY: This proposal moves the requirements for outlet boxes and box systems used to support paddle fans from Article 422 to Article 314. This was a request of the National Electrical Manufacturers Association (NEMA).

2005 CODE: "**(D) Boxes at Ceiling-Suspended (Paddle) Fan Outlets.** Outlet boxes or outlet box systems used as the sole support of a ceiling-suspended (paddle) fan shall be listed, shall be marked by their manufacturer as suitable for this purpose, and shall not support ceiling-suspended (paddle) fans that weigh more than 32 kg (70 lb). For outlet boxes or outlet box systems designed to support ceiling-suspended (paddle) fans that weigh more than 16 kg (35 lb), the required marking shall include the maximum weight to be supported."

CHANGE SIGNIFICANCE: The result of this proposal was to move the requirements for using outlet boxes for the support of ceiling-suspended (paddle) fans from Article 422 to Article 314. NEMA maintained that the change eliminates the need to refer to two sections in the Code to determine if and what size ceiling-suspended (paddle) fans are permitted to be supported by an outlet box.

314.27(D)

Boxes at Ceiling-Suspended (Paddle) Fan Outlets
NEC page 166

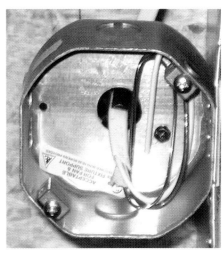

FIGURE 3.19

314.30 (New)

Handhole Enclosures
NEC page 167

CHANGE TYPE: New (Proposal 9-68a; Comments 9-111 and 9-112)

CHANGE SUMMARY: This section adds rules on the installation of handhole enclosures, including requirements for size, wiring entries, enclosures without bottoms and covers.

2005 CODE: "**314.30 Handhole Enclosures.** Handhole enclosures shall be designed and installed to withstand all loads likely to be imposed.

FPN: See ANSI/SCTE 77-2002, *Specification for Underground Enclosure Integrity*, for additional information on deliberate and non-deliberate traffic loading that can be expected to bear on underground enclosures.

(A) Size. Handhole enclosures shall be sized in accordance with 314.28(A) for conductors operating at 600 volts or below, and in accordance with 314.71 for conductors operating at over 600 volts. For handhole enclosures without bottoms where the provisions of 314.28(A)(2), Exception, or 314.71(B)(1), Exception No. 1, apply, the measurement to the removable cover shall be taken from the end of the conduit or cable assembly.

(B) Wiring Entries. Underground raceways and cable assemblies entering a handhole enclosure shall extend into the enclosure, but they shall not be required to be mechanically connected to the enclosure.

(C) Handhole Enclosures Without Bottoms. Where handhole enclosures without bottoms are installed, all enclosed conductors and any splices or terminations, if present, shall be listed as suitable for wet locations.

(D) Covers. Handhole enclosure covers shall have an identifying mark or logo that prominently identifies the function of the enclosure, such as "electric." Handhole enclosure covers shall require the use of

FIGURE 3.20

tools to open, or they shall weigh over 45 kg (100 lb). Metal covers and other exposed conductive surfaces shall be bonded in accordance with 250.96(A)."

CHANGE SIGNIFICANCE: Handhole enclosures are most comparable to pull and junction boxes, and the basic requirements are located in Article 314. This new section centralizes the field requirements in one place for user friendliness. In their proposal for the new rules, the Code Panel provides little substantiation. A corresponding definition of *handhole enclosure* has been added to Article 100.

Handhole enclosures are required to be designed and installed to withstand all loads likely to be imposed. Consideration should be given to the location of handholes, particularly to the type of traffic exposure as well as to the type of traffic that is likely to be experienced, such as pedestrian or vehicle traffic. No specific guidance is provided in the Code on determining the strength of the enclosure in terms of exposure to pedestrian vehicles.

A Fine Print Note referring to ANSI/SCTE 77-2002 is provided. ANSI/SCTE 77 more fully explores the level of loading between pedestrian and truck traffic. This allows the end user to more closely match enclosure to application and save money while maintaining safety. ANSI/SCTE 77 is produced by the Society of Cable Telecommunication Engineers. The ANSI Web site refers to the standard: "SCTE CMS WG4-0001 covers conformance test and requirements for the integrity of grade level enclosures containing telecommunication or other low voltage apparatus that may be exposed to the public. The purpose of this standard is to describe the requirements for a comprehensive integrity system for grade level enclosures providing long installation like [life] and minimal maintenance."

The reference to 314.28(A) for sizing requires the enclosure to be sized, for conductors 4 AWG or larger, based on the size of the raceway and whether the raceways pass straight through or are an angle or U pull. Since the raceways are not required to be connected to the walls of the enclosure, it is possible to keep the conduit ends closer together than if they connected to a traditional pull box. For example, if a 2-inch conduit will pass through the handhole, 314.28(A)(1) requires the width of the enclosure to be eight times the 2-inch conduit size, or 16 inches. The conduits can then extend any distance inside the enclosure. (This may present a hardship for the installer when pulling conductors into the enclosure and violate the spirit of 314.28(A)(1). Other rules of the Code, such as installing bushings on conduit ends and using splicing methods suitable for wet locations, apply.)

A handhole cover is required to have an identifying mark or logo, such as "electric." Tools must be required to open the covers, or the covers must weigh over 100 pounds—both of these requirements should ensure that the covers remain in place.

320.10 and 320.12

Armored Cable: Type AC, Uses Permitted and Uses Not Permitted

NEC page 169

CHANGE TYPE: Revision (Proposals 7-8, 7-12, and 7-13; Comments 7-14a and 7-19a)

CHANGE SUMMARY: The sections on Uses Permitted and Uses Not Permitted have been revised for Type AC, Type MC, and other cable wiring methods.

2005 CODE: "**320.10 Uses Permitted.** ~~Where not subject to physical damage,~~ Type AC cable shall be permitted as follows:

(1) In both exposed and concealed work

(2) In cable trays ~~where identified for such use~~

(3) In dry locations

(4) Embedded in plaster finish on brick or other masonry, except in damp or wet locations

(5) To be run or fished in the air voids of masonry block or tile walls where such walls are not exposed or subject to excessive moisture or dampness

FPN: The "Uses Permitted" is not an all-inclusive list."

"**320.12 Uses Not Permitted.** Type AC cable shall not be used as follows:

(1) <u>Where subject to physical damage</u>

~~In theater and similar locations, except where permitted in 518.4~~

(2) <u>In damp or wet locations</u>

~~In motion picture studios~~

FIGURE 3.21

(3) <u>In air voids of masonry block or tile walls where such walls are</u> <u>exposed or subject to excessive moisture or dampness.</u>
~~In hazardous (classified) locations except where permitted in~~
~~a. 501.4(B), Exception~~
~~b. 502.4(B), Exception No. 1~~
~~c. 504.20~~

(4) Where exposed to corrosive fumes or vapors

(5) <u>Embedded in plaster finish on brick or other masonry in damp</u> <u>or wet locations</u>
~~In storage battery rooms~~

~~(6)~~ ~~In hoistways, or on elevators or escalators, except where per-~~ ~~mitted in 620.21~~

~~(7)~~ ~~In commercial garages where prohibited in 511.4 and 511.7~~"

CHANGE SIGNIFICANCE: The NEC Technical Correlating Committee identified concerns with using the terms *uses permitted* and *uses not permitted*. Panel 7 agreed with the concept of deleting the previous .10 sections on uses permitted and expanding the previous .12 sections on uses not permitted in all articles that were Panel 7's responsibility. This action was reversed at the comment stage and revisions were instead made to both 320.10, "Uses Permitted," and 320.12, "Uses Not Permitted."

The most significant changes were made to 320.12. All of the previous uses not permitted were deleted, except for (4), to incorporate the concept of the organization of the Code in 90.3. This provides that Chapters 1 through 4 of the Code apply generally and Chapters 5, 6, and 7 can supplement or modify the rules in Chapters 1 through 4. For example, 320.12(2) previously stated "In motion picture studios" to indicate that Type AC cable was not permitted in motion picture theaters. However, 530.11 provides, "The permanent wiring shall be Type MC cable, Type AC cable containing an insulated equipment grounding conductor sized in accordance with Table 250.122..." Under the revised rules, Article 320 applies generally to permitted uses; Chapters 5, 6, and 7 will supplement or modify the general rules where appropriate.

320.30

Type AC Cable, Securing and Supporting
NEC page 169

CHANGE TYPE: Revision (Proposal 7-21; Comment 7-19)

CHANGE SUMMARY: The rules on securing and supporting Type AC cables as well as Type MC cable have been revised and separated.

2005 CODE: "**320.30 Securing and Supporting.**

<u>(A) General.</u> Type AC cable shall be supported and secured by staples, cable ties, straps, hangers, or similar fittings, designed and installed so as not to damage the cable~~, at intervals not exceeding 1.4 m (4 12 ft) and within 300 mm (12 in.) of every outlet box, junction box, cabinet, or fitting~~.

<u>(B) Securing.</u> <u>Unless otherwise provided, Type AC cable shall be secured within 300 mm (12 in.) of every outlet box, junction box, cabinet, or fitting and at intervals not exceeding 1.4 m (4 ½ ft) where installed on or across framing members.</u>

<u>(C) Supporting.</u> <u>Unless otherwise provided, Type AC cable shall be supported at intervals not exceeding 1.4 m (4 ½ ft).</u>
<u>(1)</u>~~(A)~~ Horizontal Runs **~~Through Holes and Notches.~~** ~~In other than vertical runs,~~ of type AC cable installed in <u>wooden or metal framing members or similar supporting means</u> ~~accordance with 300.4~~ shall be considered supported ~~and secured~~ where such support does not exceed 1.4-m (4 ½-ft) intervals ~~and the armored cable is securely fastened in~~

FIGURE 3.22

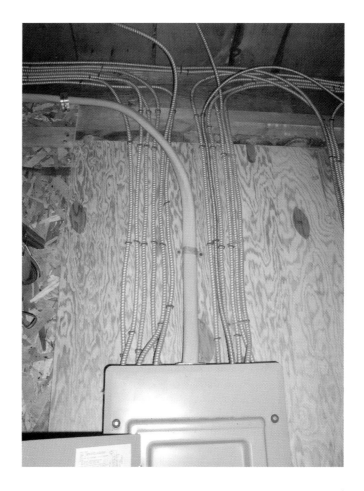

~~place by an approved means within 300 mm (12 in.) of each box, cabinet, conduit body, or other armored cable termination~~.

(D) (B) **Unsupported Cables.** Type AC cable shall be permitted to be unsupported where the cable complies with any of the following:

(1) Is fished between access points through concealed spaces in finished buildings or structures and supporting is impracticable

(2) Is not more than 600 mm (2 ft) in length at terminals where flexibility is necessary

(3) Is not more than 1.8 m (6 ft) in length from the last point of cable support to the point of connection to a luminaire(s) [lighting fixture(s)] or other ~~piece of~~ electrical equipment and the cable and point of connection are within an accessible ceiling. For the purposes of this section, Type AC cable fittings shall be permitted as a means of cable support."

CHANGE SIGNIFICANCE: As can be seen above, this section has been significantly revised and reorganized since the 2002 NEC. Significant changes include creating (A) General, (B) Securing, (C) Supporting, and (D) Unsupported Cables. The previous requirements on securing the cable every $4^1/_2$ feet and within 12 inches of every outlet box, junction box, cabinet, or fitting remain unchanged.

The rules for supporting Type AC cable clarify that the horizontal runs of cable in wooden or metal framing members are considered supported where the support does not exceed $4^1/_2$-foot-intervals. Rules on unsupported cables have also been revised to improve the organization of the material, and now include the statement "Type AC Cable fittings shall be permitted as a means of cable support." This answers the question as to whether a fitting both secures and supports the cable at its terminations.

Similar changes were made for securing and supporting Type MC cable in 330.30.

320.80(A)

Type AC Cable, Ampacity, Thermal Insulation

NEC page 170

CHANGE TYPE: Revision (Proposal 7-22)

CHANGE SUMMARY: Changes have been made to clarify that the 90°C conductor allowable ampacity is permitted to be used for derating purposes.

2005 CODE: "**(A) Thermal Insulation.** Armored cable installed in thermal insulation shall have conductors rated at 90°C (194°F). The ampacity of cable installed in these applications shall be that of 60°C (140°F) conductors. <u>The 90°C (194°F) rating shall be permitted to be used for ampacity derating purposes, provided the final derated ampacity does not exceed that for a 60°C (140°F) rated conductor.</u>"

CHANGE SIGNIFICANCE: The added sentence brings this section into harmony with 334.80 on ampacity derating of Type NM cable.

As a related example, 12-2 Type AC cables are bundled together longer than 24 inches. Section 310.15(B)(2)(a), Exception No. 5, requires that Type AC cable be derated to 60 percent of its allowable ampacity where more than 20 current-carrying conductors are bundled. The allowable ampacity of 12 AWG copper wire in the 90°C column of Table 310.16 is 30 amperes; 30 × 6 = 18 amperes. The rules in 240.4(B) determine whether it is permissible to round up to the next standard overcurrent device of 20 amperes, or if it is necessary to round down to 15 amperes.

FIGURE 3.23

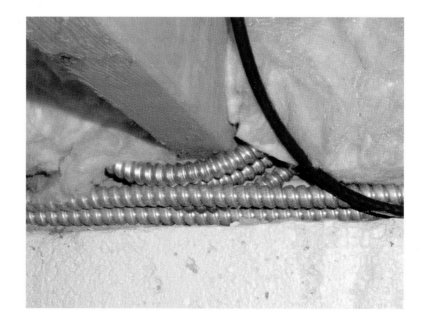

CHANGE TYPE: Revision (Proposal 7-115)

CHANGE SUMMARY: Revisions to this section are similar to those proposed by the usability task group of the NEC Technical Correlating Committee to the .12 sections of most of the cable wiring method articles. This group also proposed to eliminate the .10 sections on uses permitted. The concept being espoused was that installations are permitted if not specifically prohibited.

2005 CODE: "**334.12 Uses Not Permitted.**

(A) Types NM, NMC, and NMS. Type NM, NMC, and NMS cables shall not be permitted ~~used~~ as follows:

(1) In any dwelling or structure not specifically permitted in 334.10(1), (2), and (3) ~~As open runs in dropped or suspended ceilings in other than one- and two-family and multifamily dwellings~~.

(2) <u>Exposed in dropped or suspended ceilings in other than one- and two-family and multifamily dwellings</u>.

(3) ~~(2)~~ As service-entrance cable.

(4) ~~(3)~~ In commercial garages having hazardous (classified) locations as defined in 511.3.

(5) ~~(4)~~ In theaters and similar locations, except where permitted in 518.4(B).

(6) ~~(5)~~ In motion picture studios.

(7) ~~(6)~~ In storage battery rooms.

(8) ~~(7)~~ In hoistways or on elevators or escalators.

334.12(A)

Type NM Cable, Uses Not Permitted
NEC page 179

FIGURE 3.24

(9)   ~~(8)~~ Embedded in poured cement, concrete, or aggregate.

(10)   ~~(9)~~ In hazardous (classified) locations, except where permitted by the following:

    a. 501.10(B)(3)

    b. 502.10(B)(3)

    c. 504.20

(B) ~~(10)~~ **Types NM and NMS.** Type NM and NMS cables shall not be used under the following conditions or in the following locations:

(1) Where exposed to corrosive fumes or vapors

(2) Where embedded in masonry, concrete, adobe, fill, or plaster

(3) In a shallow chase in masonry, concrete, or adobe and covered with plaster, adobe, or similar finish

(4) Where exposed or subject to excessive moisture or dampness"

CHANGE SIGNIFICANCE: The Panel reversed itself on the .10 "Uses Permitted" section of this and several other cable wiring method articles. Subsequent modifications were made to the .12 sections of several articles. Some changes were editorial while others were more substantive.

Due to comments from some sectors of the electrical industry, it was determined that deleting the "Uses Permitted" sections and maintaining only the "Uses Not Permitted" sections was not in the best interest of the Code. Doing so would likely cause confusion and interpretation problems for inspectors, manufacturers, electricians, and others in the electrical industry.

334.15(B)

Type NM Cable, Protection From Physical Damage
NEC page 179

CHANGE TYPE: Revision (Proposal 7-132; Comment 7-115)

CHANGE SUMMARY: The rules on protection of Type NM cable have been revised and simplified. Some previously accepted methods of protecting Type NM cable against physical damage have been deleted.

2005 CODE: "**(B) Protection from Physical Damage.** Cable shall be protected from physical damage where necessary by <u>rigid metal</u> conduit, <u>intermediate metal conduit</u>, electrical metallic tubing, Schedule 80 PVC rigid non-metallic conduit, ~~pipe, guard strips, listed surface metal or nonmetallic raceway~~, or other <u>approved</u> means. Where passing through a floor, the cable shall be enclosed in rigid metal conduit, intermediate metal conduit, electrical metallic tubing, Schedule 80 PVC rigid non-metallic conduit, ~~listed surface metal or nonmetallic raceway, or other metal pipe~~ or other approved means extending at least 150 mm (6 in.) above the floor.

Where type NMC cable is installed in shallow chases in masonry, concrete, or adobe, the cable shall be protected against nails or screws by a steel plate at least 1.59 mm ($^{1}/_{16}$ in) thick and covered with plaster, adobe, or similar finish."

CHANGE SIGNIFICANCE: Note that in this revision pipe, guard strips, listed surface metal, and non-metallic raceways are not listed for areas of physical damage and should not be referenced. These methods will not provide the physical strength required to protect the NM cable in an area that has been determined as an area of physical damage.

FIGURE 3.25

334.30

Type NM Cable, Securing and Supporting
NEC page 180

CHANGE TYPE: Revision (Proposal 7-144)

CHANGE SUMMARY: This proposal adds a new paragraph: "Sections of cable protected from physical damage by a raceway shall not be required to be secured within the raceway."

2005 CODE: "**334.30 Securing and Supporting.** Nonmetallic-sheathed cable shall be secured by staples, cable ties, straps, hangers, or similar fittings designed and installed so as not to damage the cable, at intervals not exceeding 1.4 m (4$^1/_2$ ft) and within 300 mm (12 in.) of every outlet box, junction box, cabinet, box, or fitting. Flat cables shall not be stapled on edge.

<u>Sections of cable protected from physical damage by raceway shall not be required to be secured within the raceway.</u>"

CHANGE SIGNIFICANCE: This additional language resolves the issue of how many inspectors were not allowing the practice of sleeving Type NM cables because the cable was not secured within 12 inches of the box, a general requirement of 314.30.

Sections 334.15(B) and (C) have long permitted Type NM cable to be protected or sleeved by conduit or EMT. Specific requirements are provided for using a conduit or tubing on a wall of an unfinished basement as a sleeve for installation of Type NM cable to an outlet or switch box. Where this method is used, it is not practical to secure the Type NM cable to the box nor to secure the cable within 12 inches of the box. Type NM cable should be secured within 12 inches from where it enters the sleeve, and the cable must be protected from any sharp edge by a non-metallic bushing or adapter at the point of transition.

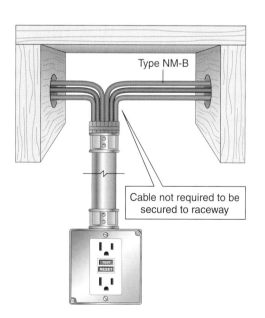

FIGURE 3.26

CHANGE TYPE: Revision (Proposal 7-150a)

CHANGE SUMMARY: Since tests show excessive temperatures occur in the confined space, this proposal adds a new paragraph to 334.80, extending the derating requirements to three or more Type NM cables that are fire- or draft-stopped where passing through wooden framing.

2005 CODE: "**334.80 Ampacity.** The ampacity of Type NM, NMC, and NMS cable shall be determined in accordance with 310.15. The ampacity shall be in accordance with the 60°C (140°F) conductor temperature rating. The 90°C (194°F) rating shall be permitted to be used for ampacity derating purposes, provided the final derated ampacity does not exceed that for a 60°C (140°F) rated conductor. The ampacity of Type NM, NMC, and NMS cable installed in cable tray shall be determined in accordance with 392.11.

Where more than two NM cables containing two or more current-carrying conductors are bundled together and pass through wood framing that is to be fire- or draft-stopped using thermal insulation or sealing foam, the allowable ampacity of each conductor shall be adjusted in accordance with Table 310.15(B)(2)(a)."

CHANGE SIGNIFICANCE: The proposal to add the requirement for derating Type NM cable in the application identified was acted upon by CMP-6 as Proposal 6-31. The Code Panel rejected the proposal to 310.15(B)(2)(a) and provided the following Panel statement: "The Panel agrees with the intent of the Proposal. However, this material is more appropriately addressed in 334.80 since the Proposal only applies to one type of cable, and Code-Making Panel 6 covers all wiring methods.

334.80
Type NM Cable, Ampacity
NEC page 180

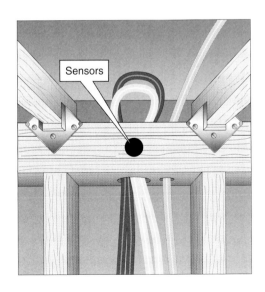

FIGURE 3.27

Therefore, Code-Making Panel 6 has forwarded this Proposal to Code-Making Panel 7 for action."

The substantiation for Proposal 6-31 read "Recent experimentation shows the possibility of dangerous conditions when loaded circuits are brought into close proximity to each other inside a fire- or draft-stop, where the ability to dissipate heat is extremely limited. Cable temperatures well in excess of their 90°C rating were encountered, with no overcurrent protection present for these conditions. Results indicate that immediate adjustments should be made to the NEC to apply at least to the specific case represented by the experiment. Such a proposal is being made, with a supplemental report offered as technical support."

The documentation supplied showed that when the conductors that were bundled carried current, dangerous elevated temperatures caused degradation of the insulation on the conductors. Twelve AWG circuit conductors had been loaded to 80 percent (16 amperes) for several hours when they reached their highest temperature of 233°F, nearly 40°F higher than the allowable limit for 90°C (194°F) cable. Obviously, this could result in these locations becoming fire initiation sources.

The Panel responded by requiring that where more than two NM cables containing two or more current-carrying conductors are bundled together and pass through wood framing that is to be fire- or draft-stopped using thermal insulation or sealing foam, the allowable ampacity of each conductor shall be adjusted in accordance with Table 310.15(B)(2)(a). Other cables, such as Type AC or MC, are not required to be derated unless they are bundled for more than 24 inches.

CHANGE TYPE: Revision (Proposal 8-9)

CHANGE SUMMARY: A new paragraph has been added to permit cables to be installed in IMC where not prohibited in respect to the cable article. Conduit fill has to comply with Table 1 of Chapter 9. This corrects the situation where CMP-7 allowed cables to be installed in conduit and tubing in the 1999 cycle, yet the raceway articles did not have a corresponding provision.

2005 CODE: "**342.22 Number of Conductors.** The number of conductors shall not exceed that permitted by the percentage fill specified in Table 1, Chapter 9.

Cables shall be permitted to be installed where such use is not prohibited permitted by the respective cable articles. The number of cables shall not exceed the allowable percentage fill specified in Table 1, Chapter 9."

CHANGE SIGNIFICANCE: The second paragraph of this section was added during the 2002 NEC cycle. The substantiation for adding this provision was that "The proposed language clarifies that cables, where permitted elsewhere in the Code, are allowed to be used in a raceway." But cable articles are structured so that installation in raceways is not prohibited. This revision will correlate the desired clarification with the cable articles.

Table 1 of Chapter 9 permits a conductor fill 53 percent for one conductor, 31 percent for two conductors, and 40 percent for over two conductors. Table note 5 indicates that for conductors not included in Chapter 9, such as multiconductor cables, the actual dimensions shall

342.22

Intermediate Metal Conduit, Number of Conductors
NEC page 184

FIGURE 3.28

be used. One method of being able to use the tables in Chapter 9 without calculating the area of the cable is to measure the diameter (width) of the cable, then scan through Table 5 to find a cable with the same approximate diameter. That row of the table then gives the approximate area in square inches of the cable. This square-inch area can then be used in Table 4 of Chapter 9 to determine the minimum size conduit or tubing required for the total square-inch area of the cables to be installed. Note also that Note 9 to Table 1 of Chapter 9 states "For cables with elliptical cross sections, the cross-sectional area calculation shall be based on using the major diameter of the ellipse as a circle diameter." This rule applies to rectangular-shaped cables such as 12-2 or 10-2/WG cables. The width of the cable is used as the diameter of a circle for purposes of determining the maximum number of cables permitted in the conduit or tubing.

Identical changes have been made to other circular raceway articles including 344.22, for rigid metal conduit; 348.22, for flexible metal conduit; 350.22, for liquidtight flexible metal conduit; 352.22, for rigid non-metallic conduit; 353.22, for high density polyethylene conduit; 356.22, for liquidtight flexible non-metallic conduit; 358.22, for electrical metallic tubing; 360.22, for flexible metallic tubing; and 362.22, for electrical non-metallic tubing.

CHANGE TYPE: Revision (Proposal 8-10)

CHANGE SUMMARY: Table 344.24, which gives the minimum radius of conduit and tubing field bends, has been moved to become Table 2 of Chapter 9.

2005 CODE: "**344.24 Bends—How Made.** Bends of RMC shall be made so that the conduit will not be damaged and so that the internal diameter of the conduit will not be effectively reduced. The radius of the curve of any field bend to the centerline of the conduit shall not be less than indicated in Table 2, Chapter 9 ~~344.24~~."

CHANGE SIGNIFICANCE: The previous Table 344.24 has been the resource for the minimum bending radius for circular raceways that are bent in the field for many Code editions. Other articles for conduit and tubing have referred to this table rather than having a duplicate table.

This proposal intended to move the table to Article 300 so that it would appear to have more universal application. The Code Panel instead moved Table 344.24 to Chapter 9. The Technical Correlating Committee agreed with the action and directed that it become Table 2 of Chapter 9 with a title of "Radius of Conduit and Tubing Bends."

344.24

Rigid Metal Conduit, Bends—How Made
NEC page 186

FIGURE 3.29

Conduit or Tubing Size		One Shot and Full Shoe Benders		Other Bends	
Metric Designator	Trade Size	mm	in.	mm	in.
16	$1/2$	101.6	4	101.6	4
21	$3/4$	114.3	$4^1/_2$	127	5
27	1	146.05	$5^3/_4$	152.4	6
35	$1^1/_4$	184.15	$7^1/_4$	203.2	8
41	$1^1/_2$	209.55	$8^1/_4$	254	10
53	2	241.3	$9^1/_2$	304.8	12
63	$2^1/_2$	266.7	$10^1/_2$	381	15
78	3	330.2	13	457.2	18
91	$3^1/_2$	381	15	533.4	21
103	4	406.4	16	609.6	24
129	5	609.6	24	762	30
155	6	762	30	914.4	36

348.30(A) Exception No. 2

Flexible Metal Conduit, Securely Fastened

NEC page 188

CHANGE TYPE: Revision (Proposal 8-43)

CHANGE SUMMARY: Section 348.30 covers how to secure and support flexible metal conduit (FMC). Changes have been made to Exception No. 2 and a new Exception No. 4 has been added.

2005 CODE: "**(A) Securely Fastened.** FMC shall be securely fastened in place by an approved means within 300 mm (12 in.) of each box, cabinet, conduit body, or other conduit termination and shall be supported and secured at intervals not to exceed 1.4 m (4½ ft).

Exception No. 1: Where FMC is fished

Exception No. 2: At terminals where flexibility is required, lengths shall not exceed the following ~~Lengths not exceeding~~:

(1) 900 mm (3 ft) for metric designators 16 through 35 (trade sizes ½ through 1¼) ~~at terminals where flexibility is required.~~

(2) 1200 mm (4 ft) for metric designators 41 through 53 (trade sizes 1½ through 2)

(3) 1500 mm (5 ft) for metric designators 63 (trade sizes 2½) and larger

Exception No. 3: Lengths not exceeding 1.8 m (6 ft) from a luminaire (light fixture) terminal connection for tap connections to luminaires (lighting fixtures) as permitted in 410.67(C).

Exception No. 4: Lengths not exceeding 1.8 m (6 ft) from the last point where the raceway is securely fastened for connections within an accessible ceiling to luminaire(s) [lighting fixture(s)] or other equipment."

CHANGE SIGNIFICANCE: The general rule is that FMC is required to be secured every 4½ feet and within 12 inches of terminations. Exception No. 2 previously permitted up to 3 feet to be unsupported

FIGURE 3.30

"at terminals where flexibility is required." With this change, the 3-foot distance applies only to trade sizes $^1/_2$ through $1^1/_4$; a 4-foot distance applies to trade sizes $1^1/_2$ and 2, and a 5-foot distance applies to trade sizes $2^1/_2$ and larger.

No explanation is provided for use of the phrase "at terminals where flexibility is required." Perhaps this means the same as the word *termination* in the main body of the section or "where the FMC ends or is terminated." The use of the word *terminals* should not imply "where the FMC ends at wire terminals."

Exception No. 4 is new and adds a permission to use unsupported FMC up to 6 ft "for connections within accessible ceilings to luminaires or other equipment." This change parallels an identical permission added to the 2002 NEC for Type AC cable and Type MC cable. The 6-foot measurement is from the last point the FMC is "securely fastened," such as to a framing member or other support means, like a strut or beam. It is not required that the 6-foot measurement begin at a junction box.

A similar change was made by Proposal 8-64 for liquidtight flexible metal conduit [350.30(a)]; in Proposal 8-114 for liquidtight flexible non-metallic conduit (356.30); and in Proposal 8-162 for electrical non-metallic tubing [362.30(A)].

352.12(E)

Rigid Nonmetallic Conduit, Uses Not Permitted, Insulation Temperature Limitations

NEC page 191

CHANGE TYPE: Revision (Proposal 8-78)

CHANGE SUMMARY: Changes have been made to clarify the accepted use of rigid non-metallic conduit (RNC) when the temperature rating of cables installed in RNC exceeds the temperature rating of the conduit. Conductors with a higher temperature rating than the RNC are permitted to be installed in the conduit when the cables are *operated* at a temperature that does not exceed the temperature rating of the conduit.

2005 CODE: "**(E) Insulation Temperature Limitations.** For conductors <u>or cables</u> ~~whose insulation~~ <u>operating at a</u> temperature <u>higher than</u> ~~limitations would exceed those for which~~ the <u>RNC</u> ~~conduit is~~ listed <u>operating temperature rating.</u>

    <u>Exception: Conductors or cables rated at a temperature higher than the RNC listed temperature rating shall be permitted to be installed in RNC, provided they are not operated at a temperature higher than the RNC listed temperature rating.</u>"

CHANGE SIGNIFICANCE: The substantiation for this proposal points out that there are numerous conductors and multiconductor cables that are rated at a higher temperature than the RNC listed temperature rating. The new exception will permit higher rated conductors or cables to be installed in RNC provided they are not operated at a temperature higher than the RNC listed temperature rating. Thus, the temperature rating of the RNC will not be exceeded, equivalent safety will be provided, and other products will not be prohibited from being installed in RNC.

FIGURE 3.31

The Underwriters Laboratories 2003 General Information Directory contains the following information on rigid non-metallic schedule 40 and schedule 80 PVC conduit (DZYR): "Unless marked for higher temperature, rigid nonmetallic conduit is intended for use with wires rated 75°C or less including where it is encased in concrete within buildings and where ambient temperature is 50°C (122°F) or less. Where encased in concrete and trenches outside a building it is suitable for use with wires rated 90°C or less."

Article 353

High Density Polyethylene Conduit: Type HDPE Conduit

NEC page 193

CHANGE TYPE: New (Proposal 8-96)

CHANGE SUMMARY: Discussion of high density polyethylene conduit (Type HDPE) has been added to the NEC as new Article 353. HDPE is permitted in "discrete lengths or in continuous lengths from a reel" and in "direct burial installations in earth or concrete." It is not permitted where exposed or within a building (like black poly water pipe). This was previously included in Article 352, but, due to the installation restrictions, is now located in this new article.

2005 CODE: "**ARTICLE 353 High Density Polyethylene Conduit: Type HDPE Conduit**

I. General

353.1 Scope. This article covers the use, installation, and construction specifications for high density polyethylene (HDPE) conduit and associated fittings.

353.2 Definition.

High Density Polyethylene (HDPE) conduit. A non-metallic raceway of circular cross section, with associated couplings, connectors, and fittings for the installation of electrical conductors.

353.6 Listing Requirements. HDPE conduit and associated fittings shall be listed.

II. Installation

353.10 Uses Permitted. The use of HDPE conduit shall be permitted under the following conditions:

(1) In discrete lengths or in continuous lengths from a reel

(2) In locations subject to severe corrosive influences as covered in 300.6 and where subject to chemicals for which the conduit is listed

FIGURE 3.32

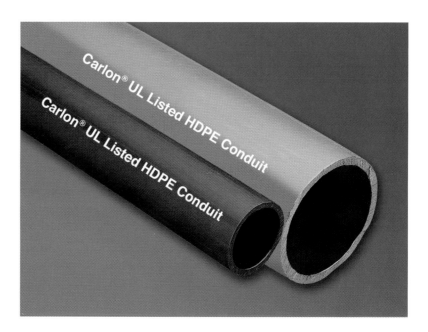

(3) In cinder fill

(4) In direct burial installations in earth or concrete

FPN: Refer to 300.5 and 300.50 for underground installations.

353.12 Uses Not Permitted. HDPE conduit shall not be used under the following conditions:

(1) Where exposed

(2) Within a building

(3) In hazardous (classified) locations, except as permitted in 504.20

(4) Where subject to ambient temperatures in excess of 50°C (122°F) unless listed otherwise

(5) For conductors or cables operating at a temperature higher than the HDPE conduit operating temperature rating

Exception: Conductors or cables rated at a temperature higher than the HDPE conduit listed temperature rating shall be permitted to be installed in HDPE conduit, provided they are not operated at a temperature higher than the HDPE conduit listed temperature rating."

(See the 2005 NEC for the remainder of Article 353.)

CHANGE SIGNIFICANCE: The justification for the new article included the fact that HDPE conduit is currently a listed product but has different locations where it is permitted to be installed, and thus does not fit into Article 352 on rigid non-metallic conduit. It is restricted in its uses and is sometimes substituted as a rigid non-metallic conduit for use underground. This new article will clarify the HDPE conduit installation requirements and construction specifications. (Annex C does not have to be revised since Table C10 and C10(A) already include HDPE conduit.)

HDPE conduit is covered in the Underwriters Laboratories General Information Directory under the category "Rigid Nonmetallic Underground Conduit, Plastic (EAZX)." This section contains the following information on the installation of HDPE conduit: "The conduit is intended for underground use under the following conditions as indicated on listing mark: (3) Direct burial with or without being encased in concrete (HDPE schedule 40). The conduit is intended for use in ambient temperatures of 50°C or less. Unless marked otherwise, type A and HDPE schedule 40 conduit are intended for use with wires rated 75°C or less. HDPE schedule 40 conduit, when directly buried or encased in concrete and trenches outside of buildings, may be used with wires rated 90°C or less. HDPE conduit is designed for joining by threaded couplings, drive on couplings, or butt fusing process. Instructions supplied by the solvent-type cement manufacturer describes the method of assembly and precautions to be followed."

372.17

Cellular Concrete Floor Raceways, Ampacity of Conductors

NEC page 208

CHANGE TYPE: New (Proposal 8-202)

CHANGE SUMMARY: Ampacity adjustment factors are now required for conductors installed in cellular concrete floor raceways. The same addition has been made to 374.17 for cellular metal floor raceways and to 390.17 for underfloor raceways.

2005 CODE: "**372.17 Ampacity of Conductors.** The ampacity adjustment factors, provided in 310.15(B)(2), shall apply to conductors installed in cellular concrete floor raceways."

CHANGE SIGNIFICANCE: The substantiation for this proposal indicates that relatively few electrical design engineers, electrical inspectors, plan-checkers, and electricians realize (or agree) that the ampacity adjustment factors (of Article 310) must be applied to conductors in cellular concrete floor raceways. This has resulted in many installations where such raceways are filled to 40 percent with no derating. The general belief seems to be that such derating would defeat much of the advantage of using this type of raceway. The NEC now clearly states this requirement in the text of Article 372.

FIGURE 3.33

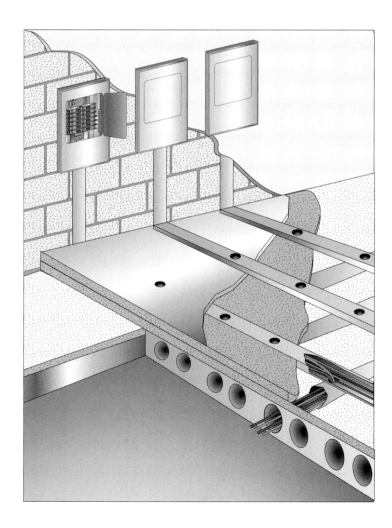

376.23(A) and (B)

Metal Wireways, Insulated Conductors
NEC page 210

CHANGE TYPE: Revision (Proposals 8-214 and 8-216)

CHANGE SUMMARY: This revision adds "one wire per terminal" to the text that was added in the 2002 NEC. This applies to deflecting conductors in a metal wireway and clarifies the proper use of Table 312.6(A). Text has also been added to introduce a rule for transposing cable sizes into raceway (conduit and tubing) sizes where cables enter a wireway, applying the pull box rule to these cables.

2005 CODE: "**376.23 Insulated Conductors.** Insulated conductors installed in a metallic wireway shall comply with 376.23(A) and (B).

(A) **Deflected Insulated Conductors.** Where insulated conductors are deflected within a metallic wireway, either at the ends or where conduits, fittings, or other raceways or cables enter or leave the metallic wireway, or where the direction of the metallic wireway is deflected greater than 30 degrees, dimensions corresponding to <u>one wire per terminal in Table </u>312.6(A) shall apply.

(B) **Metallic Wireways Used as Pull Boxes.** Where insulated conductors 4 AWG or larger are pulled through a wireway, the distance between raceway and cable entries enclosing the same conductor shall not be less than that required in 314.28(A)(1) for straight pulls and 314.28(A)(2) for angle pulls. <u>When transposing cable size into raceway size, the minimum metric designator (trade size) raceway required for the number and size of conductors in the cable shall be used.</u>"

CHANGE SIGNIFICANCE: The substantiation for this proposal maintained that the previous language led to inconsistent sizing of wireways. Because of the previous general reference to 312.6(A), users of the

FIGURE 3.34

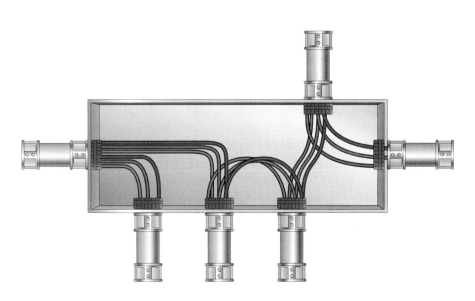

Code were applying the wiring space requirements for parallel conductors by using the *multiple*-wires-per-terminal columns in Table 312.6(A). But if the same number of conductors are routed in exactly the same manner, but not installed as parallel conductors, it is permitted to use the one-conductor-per-terminal column in Table 312.6(A).

For example, assume that three 300-kcmil conductors are installed in parallel. Table 312.6(A) indicates that a wireway 10 inches in width is required if the wireway is deflected more than 30 degrees. If three separate 300-kcmil conductors are installed in the same arrangement but are not paralleled, a 5-inch wireway is required. (The wireway must first be large enough to accommodate the number of conductors in accordance with 376.22, which permits no more than 20 percent fill.)

A rule was also added to 376.23(B) for transposing cable size into a raceway size so that the requirements in this section can be applied. The rule states that the trade size raceway required for the number and size of conductors in the cable must be used.

If the conductors in the cable are all the same size, Annex C can be used to determine the minimum raceway size. If different sizes of conductors are using the cable, Table 5 of Chapter 9 can be used to determine the conductor square-inch area; then Table 4 of Chapter 9 can be used to determine a minimum size conduit or tubing required for the conductors. After the minimum conduit size is determined in this manner, the rules in 314.28(A)(1) can be used for straight pulls (eight times the largest raceway size) and the rules in 314.28(A)(2) can be used for angle pulls (six times the maximum raceway size).

CHANGE TYPE: Revision (Proposals 8-221 and 8-223)

CHANGE SUMMARY: As with the revision to 376.23(A) and (B), this revision adds "one wire per terminal" to the text that was added to the 2002 NEC. This clarification applies to deflecting conductors in non-metallic wireways and explains the proper use of Table 312.6(A). Text has also been added to introduce a rule for transposing cable sizes into raceway (conduit and tubing) sizes where cables enter a wireway, applying the pull box rule to these cables.

2005 CODE: "**378.23 Insulated Conductors.** Insulated conductors installed in a non-metallic wireway shall comply with 378.23(A) and (B).

(A) Deflected Insulated Conductors. Where insulated conductors are deflected within a non-metallic wireway, either at the ends or where conduits, fittings, or other raceways or cables enter or leave the non-metallic wireway, or where the direction of the non-metallic wireway is deflected greater than 30 degrees, dimensions corresponding to <u>one wire per terminal</u> in Table 312.6(A) shall apply.

(B) Nonmetallic Wireways Used as Pull Boxes. Where insulated conductors 4 AWG or larger are pulled through a wireway, the distance between raceway and cable entries enclosing the same conductor shall not be less than that required in 314.28(A)(1) for straight pulls and in 314.28(A)(2) for angle pulls. <u>When transposing cable size into raceway size, the minimum metric designator (trade size) raceway required for the number and size of conductors in the cable shall be used.</u>"

FIGURE 3.35

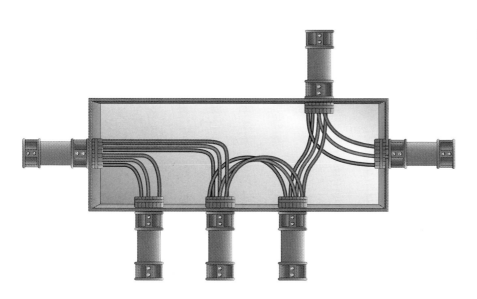

CHANGE SIGNIFICANCE: The substantiation for this proposal maintained that the previous language led to inconsistent sizing of wireways. Because of the previous general reference to 312.6(A), users of the Code were applying the wiring space requirements for parallel conductors by using the *multiple*-wires-per-terminal columns in Table 312.6(A). But if the same number of conductors are routed in exactly the same manner, but not installed as parallel conductors, it is permitted to use the one-conductor-per-terminal column in Table 312.6(A).

For example, assume that three 300-kcmil conductors are installed in parallel. Table 312.6(A) indicates that a wireway 10 inches in width is required if the wireway is deflected more than 30 degrees. If three separate 300-kcmil conductors are installed in the same arrangement but are not paralleled, a 5-inch wireway is required. (The wireway must first be large enough to accommodate the number of conductors in accordance with 378.22, which permits no more than 20 percent fill.)

A rule was also added to 378.23(B) for transposing cable size into a raceway size so that the requirements in this section can be applied. The rule states that the trade size raceway required for the number and size of conductors in the cable must be used.

If the conductors in the cable are all the same size, Annex C can be used to determine the minimum raceway size. If different sizes of conductors are using the cable, Table 5 of Chapter 9 can be used to determine the conductor square-inch area; then Table 4 of Chapter 9 can be used to determine a minimum size conduit or tubing required for the conductors. After the minimum conduit size is determined in this manner, the rules in 314.28(A)(1) can be used for straight pulls (eight times the largest raceway size) and the rules in 314.28(A)(2) can be used for angle pulls (six times the maximum raceway size).

CHANGE TYPE: Revision (Proposal 8-271)

CHANGE SUMMARY: This section covers uses not permitted for cable trays. It has now been clarified that cable trays are not permitted in certain ducts, plenums, and other air-handling spaces.

2005 CODE: "**392.4 Uses Not Permitted.** Cable tray systems shall not be used in hoistways or where subject to severe physical damage. Cable tray systems shall not be used in <u>ducts, plenums and other air-handling spaces</u> ~~environmental airspaces~~, except as permitted in 300.22, to support wiring methods recognized for use in such spaces."

CHANGE SIGNIFICANCE: The objective of this revision is to clarify the language previously used in the NEC, and to coordinate with the language used in NFPA 90A, the Standard for the Installation of Air-Conditioning and Ventilating Systems, which is the responsibility of the Technical Committee on Air Conditioning.

   As revised, it is clear that cable trays, as support systems rather than wiring methods, are permitted to support wiring methods in ducts, plenums, and other wiring spaces only when the wiring method itself is permitted to be used in that space.

392.4

Cable Trays, Uses Not Permitted
NEC page 218

FIGURE 3.36

CHAPTER 4

Equipment for General Use, Articles 400–490

400.5

Ampacities for Flexible Cords and Cables

NEC page 235

CHANGE TYPE: Revision (Proposal 6-78)

CHANGE SUMMARY: This section now requires the application of correction factors to flexible cord that is exposed to ambient temperatures higher than 30°C.

2005 CODE: "**400.5 Ampacities for Flexible Cords and Cables.**

(A) Ampacity Tables. Table 400.5(A) provides the allowable ampacities, and Table 400.5(B) provides the ampacities for flexible cords and cables with not more than three current-carrying conductors. These tables shall be used in conjunction with applicable end-use product standards to ensure selection of the proper size and type. Where the cords are used in ambient temperatures exceeding 30°C (86°F), the temperature correction factors from Table 310.16 that correspond to the temperature rating of the cord shall be applied to the ampacity from Table 400.5(B). Where If the number of current-carrying conductors exceeds three, the allowable ampacity or the ampacity of each conductor shall be reduced from the 3-conductor rating as shown in Table 400.5."

CHANGE SIGNIFICANCE: This change is intended to improve the usability of this section. Temperature ratings are not given for most of the cord types in Article 400, but listing standards state that cords are 60°C unless marked otherwise, so the temperature rating of a cord is usually easily determined by a user. However, the ampacity tables [400.5(A) and (B)] are based on the ambient temperature of 30°C (86°F), and Article 400 does not say how to correct these values for other ambient temperatures. The existing statement that the tables are to be used with "applicable end-use product standards" is not helpful to the majority of cord users, as these standards are not furnished *with* the cords. Additionally, the standards are either unavailable or are prohibitively expensive for most users, especially for those users who, in compliance with the limitations on their use, do not install large quantities of flexible cord.

FIGURE 4.1

The previous version of the section provided a requirement to correct the ampacity of the flexible cords and cables when there are more than three current-carrying conductors in the cord or cable. This section now requires a correction of the table ampacity where the flexible cord or cables are used in an ambient temperature higher than 30°C (86°F).

400.8(7)

Flexible Cord, Uses Not Permitted

NEC page 236

CHANGE TYPE: New (Proposal 6-89)

CHANGE SUMMARY: The phrase "where subject to physical damage" has been added to this section to indicate a use not permitted for flexible cords and cables.

2005 CODE: "**(7) Where subject to physical damage.**"

CHANGE SIGNIFICANCE: The substantiation for this proposal indicates that nothing in this article specifically addresses uses where a cord is likely to be damaged. Section 110.27(B) is a general requirement, but many wiring methods have a specific requirement where subject to damage.

While there is no definition for "physical damage," it is used in many locations in the Code. (As a matter of fact, a search of the 2002 NEC reveals that the term is used 185 times.) Many times, flexible cords are run across a floor or platform, on or across docks of supply watercraft, on or across the ground to supply recreational vehicles, to supply amusement rides, and for many other uses such as pendants for supply machinery. Of these uses, many are "subject to physical damage."

Section 525.20(G) does offer "Protection. Flexible cords or cables accessible to the public shall be arranged to minimize the tripping hazard and shall be permitted to be covered with nonconductive matting, provided that the matting does not constitute a greater tripping hazard than the uncovered cables." It is even permitted to bury cables for these applications. The authority having jurisdiction will no doubt consider how cords are routed to determine that they are not "subject to physical damage."

FIGURE 4.2

CHANGE TYPE: Revision (Proposal 9-91; Comment 9-122)

CHANGE SUMMARY: This revision changes the exception to allow ground-fault circuit interrupter (GFCI) protection of metal switch yokes and consequently metal faceplates where a grounding means does not exist in the box.

2005 CODE: "**404.9 Provisions for General-Use Snap Switches.**
(B) Grounding. Snap switches, including dimmer and similar control switches, shall be effectively grounded and shall provide a means to ground metal faceplates, whether or not a metal faceplate is installed. Snap switches shall be considered effectively grounded if either of the following conditions is met:

(1) The switch is mounted with metal screws to a metal box or to a non-metallic box with integral means for grounding devices.

(2) An equipment grounding conductor or equipment bonding jumper is connected to an equipment grounding termination of the snap switch.

Exception to (B): Where no grounding means exists within the snap-switch enclosure or where the wiring method does not include or provide an equipment ground, a snap switch without a grounding connection shall be permitted for replacement purposes only. A snap switch wired under the provisions of this exception and located within reach of earth, grade conducting floors, or other conducting surfaces shall be provided with a faceplate of non-conducting, non-combustible material <u>or shall be protected by a ground-fault circuit-interrupter</u>."

CHANGE SIGNIFICANCE: This proposal was unanimously rejected by CMP-9 with the statement "GFCI devices are not intended to be a substitute for effective grounding." The Panel later reversed its decision at the comment stage and accepted the proposal. The substantiation for the comment states, "The proposal amends an exception that only applies where no effective grounding means exists. Protection by

404.9(B) Exception
Provisions for General-Use Snap Switches, Grounding
NEC page 243

FIGURE 4.3

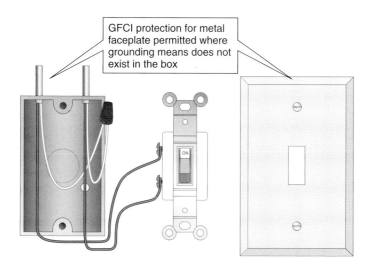

GFCI protection for metal faceplate permitted where grounding means does not exist in the box

GFCI should be allowed for switches as it is for receptacles under 406.3(D)(3)(c) and is likely to be for luminaires (see Proposal 18-69). This proposal will increase safety since a non-metallic faceplate would only cover up any fault that may occur, creating a hidden danger for anyone removing it for servicing, painting, wallpapering, etc."

A GFCI would open the circuit, thereby identifying the problem and requiring its correction. It is important to remember that this wiring method is over 40 years old and at increased risk of faulting.

One Panel member voted against the Panel action, stating, "The comment does not provide any technical substantiation to address nor refute their prior action and, accordingly, the panel should reject the comment for the originally stated reason."

406.4(A) and (B)

Receptacle Mounting
NEC page 246

CHANGE TYPE: Revision (Proposal 18-20; Comment 18-11)

CHANGE SUMMARY: Revisions have been made to this section regarding mounting of receptacles in boxes that are flush or set back from the surface.

2005 CODE: "**406.4 Receptacle Mounting.** Receptacles shall be mounted in boxes or assemblies designed for the purpose, and such boxes or assemblies shall be securely fastened in place <u>unless otherwise permitted elsewhere in this *Code*</u>.

(A) **Boxes That Are Set Back.** Receptacles mounted in boxes that are set back <u>from the finished</u> ~~of the wall~~ surface, as permitted in 314.20 shall be installed <u>such</u> ~~so~~ that the mounting yoke or strap of the receptacle is held rigidly at the surface of the wall.

(B) **Boxes That Are Flush.** Receptacles mounted in boxes that are flush with the <u>finished</u> ~~wall~~ surface or project therefrom shall be installed <u>such</u> ~~so~~ that the mounting yoke or strap of the receptacle is held rigidly against the box or ~~raised~~ box cover."

CHANGE SIGNIFICANCE: Changes were made to (A) and (B) to assure that receptacles are securely mounted to the box or are supported by the wall surface when the box is set back from the surface up to $^{1}/_{4}$ inches as permitted in 314.20. When the box is set back from the surface, either the plaster ears of the receptacle yoke must be in contact with the wall or shims must be placed between the receptacle yoke and the box to be certain the receptacle is securely mounted.

FIGURE 4.4

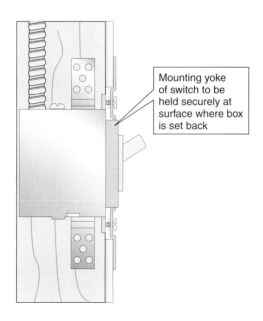

Mounting yoke of switch to be held securely at surface where box is set back

406.4(D) Exception Nos. 1 and 2

Position of Receptacle Faces

NEC page 246

CHANGE TYPE: New (Proposal 18-21; Comment 18-12a)

CHANGE SUMMARY: Two new exceptions have been added to (D) on the position of receptacle faceplates that permit receptacle kits or assemblies if the faceplate cannot be used on any other receptacle, as well as that of listed non-metallic faceplates to cover receptacle faces.

2005 CODE: "**(D) Position of Receptacle Faces.** After installation, receptacle faces shall be flush with or project from faceplates of insulating material and shall project a minimum of 0.4 mm (0.015 in.) from metal faceplates.

Exception No. 1: Listed kits or assemblies encompassing receptacles and non-metallic faceplates that cover the receptacle face, where the plate cannot be installed on any other receptacle, shall be permitted.

Exception No. 2: Listed non-metallic faceplates that cover the receptacle face to a maximum thickness of 1 mm (0.040 in.) shall be permitted."

CHANGE SIGNIFICANCE: The addition of the exceptions, in particular Exception No. 2, was a controversial and contentious issue that extended from the Report on Comments to the Technical Correlating Committee actions, was debated on the floor of the NFPA Annual Meeting, and ultimately was decided by appeal to the NFPA Standards Council.

The original CMP-18 action was on Proposal 18-21, which was accepted in principle to add a new exception to 406.4(D): "Exception: Listed kits or listed assemblies encompassing receptacles and non-metallic wall plates shall be permitted." The substantiation stated "The current wording with respect to the position of receptacle faces in faceplates of insulating material is too restrictive. The strict enforcement of this language in recently revised listing standards has resulted in the de-listing of faceplates with a very popular and effective child safety feature. These faceplates had been listed and installed for over twelve (12) years without incidence of reported problems."

FIGURE 4.5

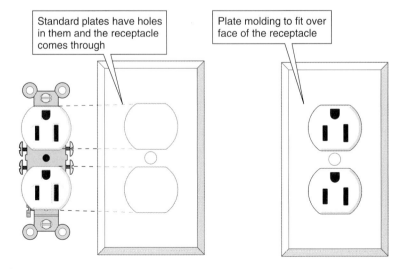

Standard plates have holes in them and the receptacle comes through

Plate molding to fit over face of the receptacle

The Code Panel stated that it "recognizes the need to permit the types of cover plates described in the substantiation. However, because there is not a dimensional standard controlling the depth of receptacle contacts below the receptacle face, caution must be exercised to assure that cover plates do not prevent attachment plug blades from properly engaging the receptacle contacts. Therefore, it would seem more likely to assure compatible products are installed if the appropriate receptacle is supplied with the cover plate or assembly with instructions and cautions against using other receptacles."

The Panel then accepted the two exceptions at the Report on Comments meeting with an extensive statement. However, following the meeting, the Technical Correlating Committee took action to hold Exception No. 2 with a statement, "The Technical Correlating Committee directs that this comment be reported as 'Accept in Part.' Exception No. 2 in the Comment will be reported as 'Hold' because the Technical Correlating Committee is concerned that not all of the safety issues related to this Exception have been addressed by the panel."

An attempt was made to raise an objection to this action on the floor of the NFPA Annual Meeting but the motion was ruled out of order. As a result, the proponent of Exception No. 2 appealed to the NFPA Standards Council on procedural grounds. The appeal was successful.

Exception No. 1 permits a kit or combination receptacle assembly "where the plate cannot be installed on any other receptacle" to be installed. Exception No. 2 permits listed non-metallic faceplates to cover the receptacle face if the faceplate is not thicker than 1 mm (0.040 in.), which allows decorator-style faceplates to be used.

406.6(A), (B), and (D)

Attachment Plugs, Cord Connectors, and Flanged Surface Devices

NEC page 247

FIGURE 4.6

CHANGE TYPE: Revisions and New (Proposals 18-29 and 18-30; Comment 18-19)

CHANGE SUMMARY: Revisions to this section include changing the title, adding titles to subsections, and adding (D), which addresses flanged surface inlets.

2005 CODE: "**406.6 Attachment Plugs, Cord Connectors, and Flanged Surface Devices.** All attachment plugs, ~~and~~ cord connectors, and flanged surface devices (inlets and outlets) shall be listed ~~for the purpose~~ and marked with the manufacturer's name or identification and voltage and ampere ratings.

(A) <u>Construction of Attachment Plugs and Cord Connectors.</u> Attachment plugs and cord connectors shall be constructed so that there are no exposed current-carrying parts except the prongs, blades, or pins. The cover for wire terminations shall be a part that is essential for the operation of an attachment plug or connector (dead-front construction).

(B) <u>Connection of Attachment Plugs.</u> Attachment plugs shall be installed so that their prongs, blades, or pins are not energized unless inserted into an energized receptacle. No receptacle shall be installed so as to require an energized attachment plug as its source of supply.

(C) **Attachment Plug Ejector Mechanisms.** Attachment plug ejector mechanisms shall not adversely affect engagement of the blades of the attachment plug with the contacts of the receptacle.

(D) <u>Flanged Surface Inlet.</u> <u>A flanged surface inlet shall be installed so that the prongs, blades, or pins are not energized unless an energized cord connector is inserted into it.</u>"

CHANGE SIGNIFICANCE: Flanged surface inlets (sometimes referred to as motor-base attachment plugs) are now addressed in this section to assure their proper use. For obvious reasons, flanged surface inlets should not be energized until a cord connector is attached (the male attachment plug prongs will be energized and could become a shock hazard).

Flanged surface inlets or outlets are referred to in other sections of the NEC, including in 410.30(C)(3) for connection of luminaires, in 551.46(A)(1) for connecting recreational vehicles, and in 552.44(B) for connecting park trailers.

CHANGE TYPE: Revision (Proposal 18-35; Comment 18-23)

CHANGE SUMMARY: Revisions have been made to the requirements for 15- and 20-ampere receptacles in wet locations and in bathtub and shower spaces. Also, a new requirement has been added for flush-mounted receptacles with faceplates.

2005 CODE: "**(B) Wet Locations.**

(1) 15- and 20-Ampere ~~Outdoor~~ Receptacles in Wet Locations. 15- and 20-ampere, 125- and 250-volt receptacles installed ~~outdoors~~ in a wet location shall have an enclosure that is weatherproof whether or not the attachment plug cap is inserted.

(2) Other Receptacles. All other receptacles installed in a wet location shall comply with (B)(2)(a) or (B)(2)(b):

(a) A receptacle installed in a wet location where the product intended to be plugged into it is not attended while in use, ~~(e.g., sprinkler system controller, landscape lighting, holiday lights, and so forth)~~ shall have an enclosure that is weatherproof with the attachment plug cap inserted or removed.

(b) A receptacle installed in a wet location where the product intended to be plugged into it will be attended while in use (e.g., portable tools~~, and so forth~~) shall have an enclosure that is weatherproof when the attachment plug is removed.

(C) Bathtub and Shower Space. ~~A~~ Receptacle_s_ shall not be installed within or directly over a bathtub or shower stall.

(D) Protection for Floor Receptacles. Standpipes of floor receptacles shall allow floor-cleaning equipment to be operated without damage to receptacles.

(E) Flush Mounting with Faceplate. The enclosure for a receptacle installed in an outlet box flush-mounted in a finished ~~on a wall~~ surface shall be made weatherproof by means of a weatherproof faceplate

406.8(B) (1), (2), (C), and (E)

Receptacles in Wet Locations

NEC page 247

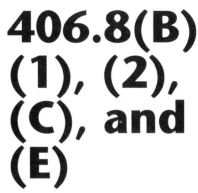

FIGURE 4.7

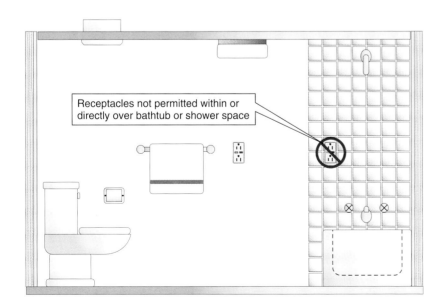

Receptacles not permitted within or directly over bathtub or shower space

assembly that provides a watertight connection between the plate and the <u>finished</u> ~~wall~~ surface."

CHANGE SIGNIFICANCE: Deleting the word "Outdoor" in the title of this section and adding "in Wet Locations" recognizes the expanded application of the section. In the text of (B)(1), deleting "outdoors" requires the enclosure to be weatherproof, whether or not the attachment plug is inserted, in any wet location, rather than just in an outdoor location. Indoor locations, such as food processing areas in supermarkets, vehicle washing areas inside car washes, and many industrial operations often require that the walls be hosed down or exposed to water, even with electrical appliances still plugged into the receptacle.

For bathtub and shower areas, the rules are expanded in (C) to prohibit receptacles from being installed directly over the units. Previously, the prohibition applied only to the area "within" a bathtub or shower area.

Some changes to (E) are editorial and some are substantive. The rules now apply to receptacles installed with a flush-mount cover in "finished surfaces," not just in walls. Also, the flush-type cover must have a watertight connection between the plate and the finished surface and not be just "weatherproof."

The term *weatherproof* is defined in Article 100 as "constructed or protected so that exposure to the weather will not interfere with successful operation." The requirement for a watertight connection is stricter than that for a weatherproof connection. *Watertight* is defined in Article 100 as "constructed so that moisture will not enter the enclosure under specified test conditions." To achieve a watertight connection, carefully follow manufacturer's installation instructions, which may require the application of a sealing compound around portions of the cover plate.

CHANGE TYPE: Revision (Proposal 9-104; Comment 9-129)

CHANGE SUMMARY: The requirements for providing a circuit directory or circuit identification have been made much more specific.

2005 CODE: "**408.4 Circuit Directory or Circuit Identification.** Every circuits and circuit modifications shall be legibly identified as to its clear, evident, and specific purpose or use. The identification shall include sufficient detail to allow each circuit to be distinguished from all others. The identification shall be included in on a circuit directory that is located on the face or inside of the panel door in the case of a panelboard, and at each switch on a switchboard."

CHANGE SIGNIFICANCE: This change provides requirements that are much more specific and exhaustive for circuit identification than required by 110.22. (However, Section 110.22 continues to include a requirement for the durability of the directory that is absent from the rules in 408.4.) Now the identification must show the "clear, evident, and specific purpose or use" of the circuit. It is obvious that the commonly found "Lights and Plugs," "Kitchen Outlets," and "Heat" will no longer be adequate. In addition, the identification must include "sufficient detail to allow each circuit to be distinguished from all others."

The level of detail to be provided will require interpretation by the authority having jurisdiction (AHJ). For example, will labels like "Kitchen Counter Receptacles Between the Sink and Refrigerator" and "Lighting for Kitchen, Dining Room, Door Outside Dining Room, and Receptacles on the South Wall of the Living Room" be required? Or will "Kitchen Receptacles" and "Lights in Kitchen and Dining Room, and Living Room Plugs" be adequate?

Keep in mind that the AHJ continues to be responsible for interpreting and enforcing the NEC. It remains to be seen whether manufacturers of panelboards will supply circuit directories with enough space to write the level of detail now required!

408.4
Circuit Directory or Circuit Identification
NEC page 249

FIGURE 4.8

408.7

Switchboards and Panelboards, Unused Openings
NEC page 249

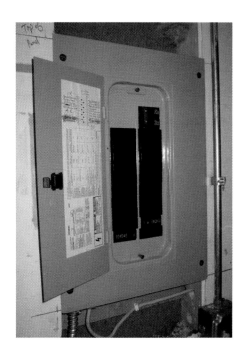

FIGURE 4.9

CHANGE TYPE:　New (Proposal 9-119; Comment 9-132)

CHANGE SUMMARY:　A new section has been added that provides requirements for closing unused openings for circuit breakers and switches similar to the rules in 110.12(A) for cable and raceway openings.

2005 CODE:　"**408.7 Unused Openings.** Unused openings for circuit breakers and switches shall be closed using identified closures, or other approved means that provide protection substantially equivalent to the wall of the enclosure."

CHANGE SIGNIFICANCE:　The change to Section 110.12 in the 2002 Code that specified unused openings as applying only to raceways and cables does not address unused breaker and switch openings in panelboards. And CMP-9 initially rejected a proposal to add the requirement to 110.12(A).

The proposal was accepted but modified for 408.7 at the Report on Comments meeting to apply specifically to circuit breaker and switch openings in panelboards and switchboards. Note that CMP-9 chose the words "identified closures" rather than "listed closures" because some closures provided by manufacturers may not be "listed."

CHANGE TYPE: New (Proposal 11-5; Comment 11-3a)

CHANGE SUMMARY: A new article has been added that covers industrial control panels intended for general use and operating at 600 volts or less. A Fine Print Note makes reference to UL 508A, the safety standard for industrial control panels.

2005 CODE:
    Article 409 Industrial Control Panels
    I. General
    409.1 Scope
    409.2 Definitions

    "**Industrial Control Panel.** An assembly of a systematic and standard arrangement of two or more components such as motor controllers, overload relays, fused disconnect switches, and circuit breakers and related control devices such as pushbutton stations, selector switches, timers, switches, control relays, and the like with associated wiring, terminal blocks, pilot lights, and similar components. The industrial control panel does not include the controlled equipment."

    409.3 Other Articles
    II. Installation
    409.20 Conductor-Minimum Size and Ampacity
    409.21 Overcurrent Protection
    409.30 Disconnecting Means
    409.60 Grounding
    III. Construction Specifications
    409.100 Enclosures
    409.102 Busbars and Conductors

Article 409

Industrial Control Panels
NEC page 252

FIGURE 4.10

409.104 Wiring Space in Industrial Control Panels
409.108 Service-Entrance Equipment
409.110 Marking

CHANGE SIGNIFICANCE: Extensive substantiation was provided for this proposal for each of the sections proposed. While the title of the article is "Industrial Control Panels," the Scope indicates that the article applies to "panels intended for general use." The new article is not limited to industrial occupancies but rather has general application; the definition of *industrial control panel* states that it is "an assembly of a systematic and standard arrangement of two or more components." Note that the scope limits the application of this article to 600 volts or less.

If the authority having jurisdiction determines that the control equipment being built is an industrial control panel, all of the requirements of the article must be complied with, including the extensive marking requirement in 409.110. No explanation is offered for what the phrase "systematic and standard arrangement of two or more components" in the definition means. Industrial control panels are often one-of-a-kind to control a specific process or operation.

While the new article does not require an industrial control panel to be listed, electrical product testing laboratories offer a program whereby "panel shops" can follow a prescribed standard for assembly of control panels using prescribed components and put a listing mark on the equipment.

Rules in other standards might apply to specific control panels, such as NFPA 79 for industrial machinery.

410.4(D)
Bathtub and Shower Areas
NEC page 254

CHANGE TYPE: Revision (Proposal 18-55; Comment 18-31)

CHANGE SUMMARY: The first sentence of (D) has been revised to expand the types of luminaires covered by the rules for bathtub and shower areas. The new last sentence requires that such a fixture be listed for a damp or wet location if it is likely to be exposed to shower spray.

2005 CODE: "**(D) Bathtub and Shower Areas.** No parts of cord-connected luminaires (fixtures); chain-, cable-, or cord-suspended luminaires (fixtures); hanging luminaires (fixtures) lighting track; pendants; or ceiling-suspended (paddle) fans shall be located within a zone measured 900 mm (3 ft) horizontally and 2.5 m (8 ft) vertically from the top of the bathtub rim or shower stall threshold. This zone is all encompassing and includes the zone directly over the tub or shower stall. Luminaires (lighting fixtures) located in this zone shall be listed for damp locations, or listed for wet locations where subject to shower spray."

CHANGE SIGNIFICANCE: Changes made to the first sentence are intended to clarify what the Code Panel meant by "hanging luminaires (fixtures)" as it now describes the methods for hanging a luminaire. This was done to clarify that the rule does not apply to luminaires that are hung on the wall, such as a wall-hung photograph.

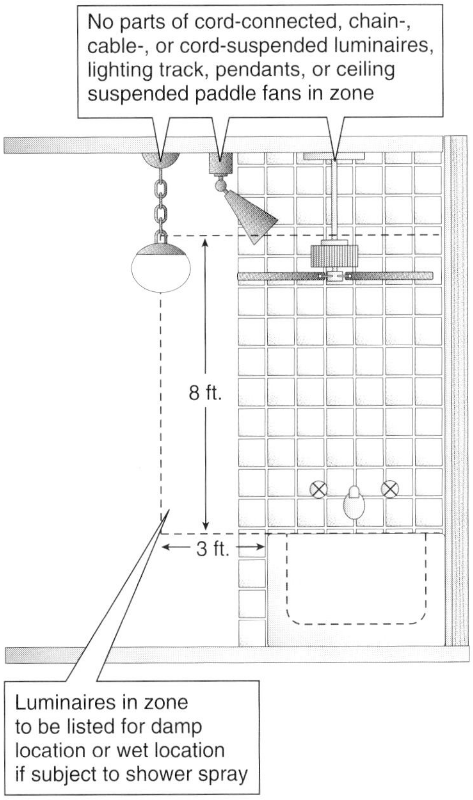

FIGURE 4.11

No parts of cord-connected, chain-, cable-, or cord-suspended luminaires, lighting track, pendants, or ceiling suspended paddle fans in zone

8 ft.

3 ft.

Luminaires in zone to be listed for damp location or wet location if subject to shower spray

The Panel made it clear in its statement on the proposal that the rules in this section are not intended to prohibit the installation of a luminaire on a wall where the luminaire is suitable for the environment. It should also be noted that there are no requirements in the NEC that luminaires in bathtub or shower areas are required to have ground-fault circuit interruption (GFCI) protection. If such a rule exists, it comes from the manufacturer of the luminaire. Where this is the case, those manufacturer installation rules must be followed in order to comply with 110.3(B).

The last sentence is new and clarifies the application of wet and damp locations as defined in Article 100 to the bathtub and shower zone. Note that this zone is "all encompassing" and extends above the rim of the bathtub for 8 feet and out 3 feet; the same "zone" applies to shower stalls. Since cord-connected luminaires, lighting track, and ceiling-suspended paddle fans are not permitted within this zone, the application of the rule is to surface-mounted luminaires within the zone. Recessed incandescent fixtures that are mounted above the zone would not be considered to be within the zone and thus are not required to comply with the new damp or wet location rules. Commonly, recessed incandescent fixtures installed above the zone are suitable for a damp location. Many are installed with a solid lens and gasketed cover to comply with the damp location requirements of the manufacturer.

CHANGE TYPE: New (Proposal 18-57; Comment 18-35)

CHANGE SUMMARY: A new section has been added that covers luminaires with certain types of lamps, which must be protected if subject to physical damage in indoor sports, mixed-use, or all-purpose facilities. The protection must be a glass or UV-attenuating plastic lens.

2005 CODE: "**410.4(E) Luminaires (Fixtures) in Indoor Sports, Mixed-Use, and All-Purpose Facilities.** Luminaires (fixtures) subject to physical damage, using a mercury vapor or metal halide lamp, installed in playing and spectator seating areas of indoor sports, mixed-use, or all-purpose facilities, shall be of the type that protects the lamp with a glass or plastic lens. Such luminaires (fixtures) shall be permitted to have an additional guard."

CHANGE SIGNIFICANCE: It has been reported that the industry continues to experience a small but persistent number of cases of damage to luminaires in sports facilities, particularly in schools, where athletic activity may result in the breakage of a mercury or metal halide lamp outer jacket. When this occurs, the arc tube can continue to operate even though the outer jacket may be cracked or no longer present. Lack of an operational jacket can result in a burning sensation around the eyes and a sunburn appearance around the cheeks and forehead. These reported symptoms of eye inflammation and skin burn are typical of overexposure to UV radiation. Fortunately, a complete lamp enclosure can reduce the possibility of damage to the outer lamp jacket.

Since this new rule includes the phrase "luminaires that are subject to physical damage," it will require a judgment on the part of the authority having jurisdiction about what mounting height or location will be outside of the "physical damage zone." Only those luminaires that are above or illuminate playing and spectator seating areas are required to be enclosed.

410.4(E)

Luminaires (Fixtures) in Indoor Sports, Mixed-Use, and All-Purpose Facilities
NEC page 254

FIGURE 4.12

410.18(B) Exception No. 2

Exposed Luminaire (Fixture) Parts Made of Insulating Material

NEC page 257

CHANGE TYPE: New (Proposal 18-69)

CHANGE SUMMARY: A new exception has been added to permit replacement luminaires to be protected by a ground-fault circuit interruption (GFCI) device if a grounding means does not exist in the outlet.

2005 CODE: "**410.18 Exposed Luminaire (Fixture) Parts.**

(A) **Exposed Conductive Parts.** Exposed metal parts shall be grounded or insulated from ground and other conducting surfaces or be inaccessible to unqualified personnel. Lamp tie wires, mounting screws, clips, and decorative bands on glass spaced at least 38 mm ($1\frac{1}{2}$ in.) from lamp terminals shall not be required to be grounded.

(B) **Made of Insulating Material.** Luminaires (fixtures) directly wired or attached to outlets supplied by a wiring method that does not provide a ready means for grounding shall be made of insulating material and shall have no exposed conductive parts.

Exception No. 1: Replacement luminaires (fixtures) shall be permitted to connect an equipment grounding conductor from the outlet in compliance with 250.130(C). The luminaire (fixture) shall then be grounded in accordance with 410.18(A).

Exception No. 2: Where no equipment grounding conductor exists at the outlet, replacement luminaires (fixtures) that are GFCI protected shall not be required to be connected to an equipment grounding conductor."

CHANGE SIGNIFICANCE: The substantiation for this proposal states that it intends to provide another method of providing protection of lighting fixtures where a grounding means does not exist at the outlet where a fixture is to be connected. The other method of providing an equipment grounding conductor external of the wiring method, as provided in Exception No. 1, was added to this section in the 2002 NEC cycle.

The new Exception No. 2 allows for a GFCI device to protect a replacement luminaire (lighting fixture) at an existing outlet (point where utilization equipment receives its supply) where an equipment grounding conductor is not present. (This new exception applies to an

FIGURE 4.13

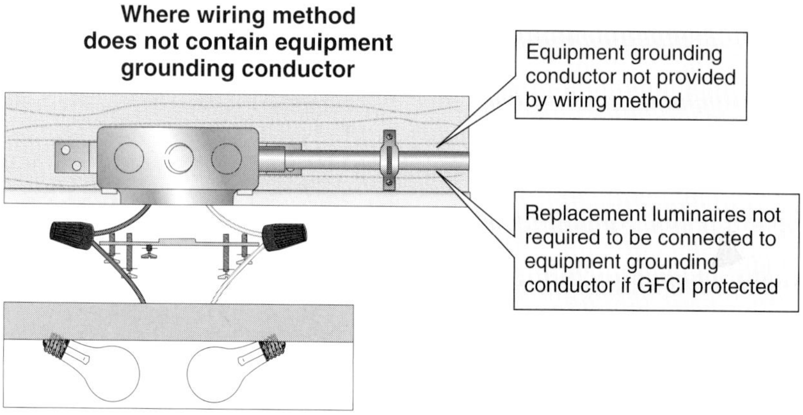

Where wiring method does not contain equipment grounding conductor

Equipment grounding conductor not provided by wiring method

Replacement luminaires not required to be connected to equipment grounding conductor if GFCI protected existing two-wire branch circuit. This older system was often wired with knob-and-tube or 2-wire Type NM cables.) The concept is similar to permitting GFCI to protect an ungrounded receptacle for replacements where no grounding means exists in the box.

It should be noted that the method provided in Exception No. 2 does not permit an equipment grounding conductor to be run from the GFCI device, since this might give the occupant the impression that the branch circuit indeed has an equipment grounding conductor in the circuit.

A GFCI of either the circuit breaker or feed-through receptacle type is permitted to be installed on the supply side of the replacement luminaire (lighting fixture).

Article 250 defers the requirements for grounding lighting fixtures to Article 410; see 250.112(J).

410.73(F) (5)

High-Intensity Discharge Luminaires (Fixtures), Metal Halide Lamp Containment

NEC page 262

CHANGE TYPE: New (Proposal 18-91)

CHANGE SUMMARY: A new subdivision has been added to place specific requirements on high-intensity discharge luminaires (fixtures) that use a metal halide lamp other than a thick-glass parabolic reflector (PAR).

2005 CODE: **"(5) Metal Halide Lamp Containment.** Luminaires (fixtures) that use a metal halide lamp other than a thick-glass parabolic reflector lamp (PAR) shall be provided with a containment barrier that encloses the lamp, or shall be provided with a physical means that only allows the use of a lamp that is Type O.

FPN: See ANSI Standard C78.387, American National Standard for Electric Lamps—Metal Halide Lamps, Methods of Measuring Characteristics."

CHANGE SIGNIFICANCE: Metal halide lamps have been identified as presenting a risk of hazardous end-of-life failure. This risk is described in papers published by the NFPA, Occupational Safety and Health Administration (OSHA), and others. The article presented a persistent loss history involving high-intensity discharge (HID) lighting, particularly with indoor industrial and warehousing applications, that has led to increased scrutiny from property underwriters and an escalated assessment by industry of HID lamps as potential ignition sources. The article went on to state, "Of the three types of HID lighting, metal halide lamps evidently display the highest potential for violent arc tube failure."

Metal halide luminaires that utilize lamp enclosures (lenses) can be evaluated and listed for their ability to contain a lamp rupture regardless of the lamp type used; however, metal halide luminaires that do not

FIGURE 4.14

have lenses can also be listed if the luminaire is labeled as not being suitable for use with certain lamp types. The lamp types that are acceptable for use in open luminaires (i.e., those that do not require lamp enclosures) are classified by lamp manufacturers as being either "O-type" or "S-type." O-type lamps have a shroud around the arc tube and are containment tested in accordance with ANSI standard C78.387, and so are rated for use in open luminaires. These lamps are frequently designed so that they fit into medium- or mogul-base lampholders that have been specifically designed for O-type lamps. S-type lamps have not been containment tested but are allowed in open luminaires provided they are oriented within 15 degrees of vertical and all of the lamp manufacturer's cautions, warnings, and instructions are followed. Depending on the specific lamp and manufacturer, this includes:

1. turning the lamp off for at least 15 minutes per week

2. group relamping at specified intervals

3. not installing the lamp over combustible material, and

4. not using the lamp in an area that is not occupied for extended periods of time.

S-type lamps do not fit into the lampholders that are designed for O-type lamps. Adhering to the lamp manufacturers' recommendations and the guidelines published in the NEMA Lighting Systems Division document #LSD 25-2002, "Best Practices for Metal Halide Lighting Systems, Plus Questions and Answers About Lamp Ruptures in Metal Halide Lighting Systems" will minimize the risks associated with S-type lamps. However, there is never assurance that the end user will adhere to these recommendations and guidelines. The potential for the end user to ignore the guidelines is a risk that can be avoided by requiring the application of the latest technology. By requiring the use of containment-tested O-type lamps and lampholders that will only accept O-type lamps, the need to rely on the maintenance practices of the end user to assure safety can be minimized.

Metal halide PAR lamps have not demonstrated a rupture risk because they have thick glass envelopes that contain arc tube particles; therefore, they should be exempt from this requirement.

410.73(G)

High-Intensity Discharge Luminaires (Fixtures), Disconnecting Means

NEC page 262

CHANGE TYPE: New (Proposal 18-93; Comment 18-52)

CHANGE SUMMARY: A new subdivision has been added to require an internal or external disconnecting means for certain fluorescent luminaires (fixtures) and other ballasted luminaires that are supplied by a multiwire branch circuit. Five exceptions are provided for certain conditions.

2005 CODE: "**410.73(G) Disconnecting Means.** In indoor locations, other than dwellings and associated accessory structures, fluorescent luminaires (fixtures) that utilize double-ended lamps and contain ballast(s) that can be serviced in place or ballasted luminaires that are supplied from multiwire branch circuits and contain ballast(s) that can be serviced in place shall have a disconnecting means either internal or external to each luminaire (fixture), to disconnect simultaneously from the source of supply all conductors of the ballast, including the grounded conductor if any. The line side terminals of the disconnecting means shall be guarded. The disconnecting means shall be located so as to be accessible to qualified persons before servicing or maintaining the ballast. This requirement shall become effective January 1, 2008.

Exception No. 1: A disconnecting means shall not be required for luminaires (fixtures) installed in hazardous (classified) location(s).

Exception No. 2: A disconnecting means shall not be required for emergency illumination required in 700.16.

Exception No. 3: For cord-and-plug connected luminaires, an accessible separable connector or an accessible plug and receptacle shall be permitted to serve as the disconnecting means.

Exception No. 4: A disconnecting means shall not be required in industrial establishments with restricted public access where conditions

FIGURE 4.15

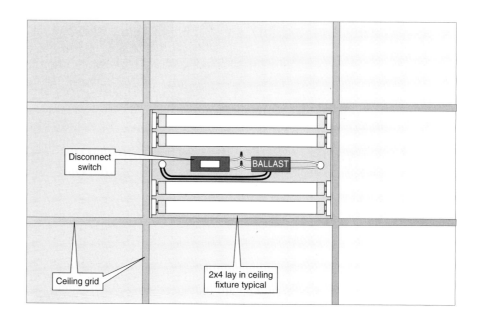

of maintenance and supervision ensure that only qualified persons service the installation by written procedures.

Exception No. 5: Where more than one luminaire is installed and supplied by other than a multiwire branch circuit, a disconnecting means shall not be required for every luminaire when the design of the installation includes locally accessible disconnects, such that the illuminated space cannot be left in total darkness.

CHANGE SIGNIFICANCE: The substantiation for this proposal stated, "Changing the ballast out in a luminaire while the circuit feeding the luminaire is energized has become a regular practice. There are several reasons for this:

1. The circuit may not be identified correctly or at all.
2. When the circuit is de-energized the room becomes dark.
3. The circuit may feed a large area and would create interruptions of the work in the area."

In a case of a terminated electrician brought before a labor board, the labor board reversed the termination based upon the fact that changing a ballast out in a luminaire while energized was standard industry practice. That decision brings this issue to a new level. As long as this is perceived as an acceptable practice, then progress in electrical safety will be slowed. Work practices alone will not change the electrical safety culture. Installation safety standards such as the NEC must provide equipment requirements that safeguard individuals who are exposed to the hazards of electricity while maintaining that equipment.

The Code Panel responded that it agreed with this proposal's concept of a local disconnecting means at the luminaire for maintenance and servicing of the ballast. The panel modified the proposal in several ways to its present form; the following points all apply to the new rule:

1. It applies in indoor locations.
2. It does not apply to dwellings and associated accessory structures.
3. It applies to fluorescent luminaires (fixtures) that utilize double-ended lamps and contain ballast(s) that can be serviced in place.
4. It applies to ballasted luminaires that are supplied from multiwire branch circuits and contain ballast(s) that can be serviced in place.
5. It applies to each luminaire (fixture) to disconnect simultaneously from the source of supply all conductors of the ballast, including the grounded conductor if any.
6. The line side terminals of the disconnecting means shall be guarded.
7. The disconnecting means shall be located so as to be accessible to qualified persons before servicing or maintaining the ballast.
8. This requirement becomes effective January 1, 2008, to allow adequate time for manufacturers and users to prepare for implementation of this requirement.

9. A disconnecting means is not required for luminaires (fixtures) installed in hazardous (classified) location(s) as there may be hazardous (classified) locations in which the operation of a disconnecting means could create a greater hazard.

10. A disconnecting means is not required for emergency illumination required in 700.16.

11. An accessible separable connector or an accessible plug and receptacle are permitted to serve as the disconnecting means for cord-and-plug connected luminaires such as permitted in 410.30(C).

12. A disconnecting means shall not be required in certain industrial establishments.

13. A disconnecting means shall not be required for every luminaire when:

- more than one luminaire is installed.
- they are supplied by other than a multiwire branch circuit.
- locally accessible disconnects are provided so that the illuminated space cannot be left in total darkness.

The required disconnecting means can be a device, a group of devices, or other means by which the conductors of a circuit can be disconnected from their source of supply. This permits the utilization of many different types of components to accomplish the requirement.

It is well known that working on energized equipment is not safe. It is also well known that if a local disconnect is not available, ballasts will be serviced while energized. Also, most ballasts are serviced from a ladder, adding the increased injury from a fall if a shock is experienced. It is common practice to install a multiwire circuit in long runs of fluorescent strip lights. If the grounded conductor in a multiwire circuit is not disconnected at the same time as the ungrounded conductor, a false sense of security could result in an unexpected shock and its consequences by the worker becoming a part of the grounded (neutral) conductor.

CHANGE TYPE: Revision (Proposal 17-15)

CHANGE SUMMARY: Changes have been made to this section to indicate that a fixed storage-type water heater is considered to be a continuous load.

2005 CODE: "**422.13 Storage-Type Water Heaters.** A ~~branch circuit supplying a~~ fixed storage-type water heater that has a capacity of 450 L (120 gal) or less shall <u>be considered a continuous load</u> ~~have a rating not less than 125 percent of the nameplate rating of the water heater~~.

FPN: For branch-circuit rating, see 422.10."

CHANGE SIGNIFICANCE: The 2002 Code text stated that the branch circuit was required to have a rating not less than 125 percent of the nameplate rating of the branch circuit. This requirement provided for voltage fluctuations above the rated nameplate voltage. However, the section stopped short of naming the load as *continuous*. This led to questions as to why a branch-circuit rating of 125 percent was required if the load was not considered to be continuous. Also, it was previously maintained that a fixed storage-type water heater is thermostatically

422.13
Storage-Type Water Heaters
NEC page 266

FIGURE 4.16

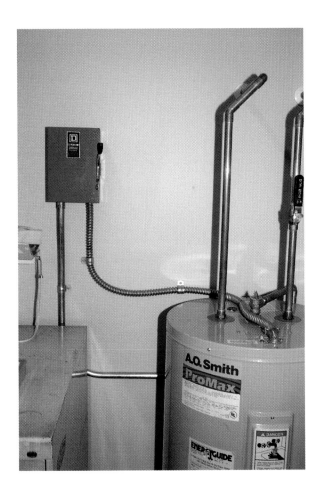

controlled and thus the load is not likely to be on for three hours or more to qualify for "continuous load" status.

One impact of this change is that now load calculations must take the continuous load requirements into account. Loads that are not continuous are considered at 100 percent (unless some demand factor applies) and are included at 125 percent for a continuous load. This change may have little impact on an average single-family dwelling but could be a significant factor in multifamily dwellings, like apartment buildings.

The Fine Print Note points to 422.10 for branch-circuit sizing. This section reads in part, "The branch-circuit rating for an appliance that is continuously loaded, other than a motor-operated appliance, shall not be less than 125 percent of the marked rating, or not less than 100 percent of the marked rating if the branch-circuit device and its assembly are listed for continuous loading and 100 percent of its rating." As can be seen, the sizing of the branch circuit for the water heater is unchanged from the previous Code edition.

It should be noted that standard circuit breakers are not suitable for operation at 100 percent of their rating. These standard circuit breakers are suitable for use at 80 percent of their rating when supplying a continuous load. This requirement correlates exactly with the 125-percent rule for continuous loads. For example, consider a 30-ampere branch circuit to be installed for a water heater. Since it is now considered a continuous load, the 30-ampere breaker is permitted to carry no more than 80 percent of its rating, or 24 amperes. When considered from the load perspective, a 24-ampere continuous load must be supplied by a branch circuit that is rated at 125 percent of the load, or 30 amperes.

CHANGE TYPE: New (Proposal 17-21)

CHANGE SUMMARY: A new subdivision has been added to specifically permit range hoods to be cord-and-plug connected. This provides for future upgrading to a combination microwave oven and range hood. The rules are similar to those provided for dishwashers and waste disposals.

2005 CODE: "**422.16(B)(4) Range Hoods**. Range hoods shall be permitted to be cord-and-plug connected with a flexible cord identified as suitable for use on range hoods in the installation instructions of the appliance manufacturer, where all of the following conditions are met:

(1) The flexible cord shall be terminated with a grounding-type attachment plug.

Exception: A listed range hood distinctly marked to identify it as protected by a system of double insulation, or its equivalent, shall not be required to be terminated with a grounding-type attachment plug.

(2) The length of the cord shall not be less than 450 mm (18 in.) and not over 900 mm (36 in.).

(3) Receptacles shall be located to avoid physical damage to the flexible cord.

(4) The receptacle shall be accessible.

(5) The receptacle shall be supplied by an individual branch circuit."

CHANGE SIGNIFICANCE: The permission to use a cord-and-plug connection provides the ability to upgrade at some time in the future to a combined microwave/range hood. When range hoods are wired with no thought for a future upgrade, the 15-ampere lighting circuit in the area

422.16(B)(4)

Flexible Cords for Specific Appliances, Range Hoods
NEC page 266

FIGURE 4.17

is typically used to power the hood fan. Most electrical inspectors will not permit the hood fan on a small-appliance branch circuit since there is no receptacle for the appliance. Earlier, hood fans were not connected with a flexible cord because such a connection would not have been seen as a permitted use for flexible cord in 400.7(8) (the hood fan is not usually removed for maintenance or repair).

It should be noted that this new rule is not initially a requirement but an optional method of making a range hood connection. However, once the decision is made to utilize this optional method of connection, all the subsequent rules must be complied with. These rules include:

1. The cord must be identified as suitable for use on range hoods. While you are not likely to find one specifically marked this way, many appliance flexible cords that are used on other appliances, such as dishwashers and waste disposers, are suitable for the purpose. 422.16(B)(1) and (2) use the phrase "with a flexible cord identified as suitable for the purpose in the installation instructions of the appliance manufacturer."

2. The flexible cord must terminate in a grounding-type attachment plug unless the hood fan is double-insulated.

3. The length of the cord must be between 18 and 36 inches.

4. Receptacles must be located to avoid physical damage and must be accessible. Note that the definition of *accessible* in Article 100 is a less strict requirement than *readily accessible* and permits the receptacle to be located at an elevated height and/or inside a cabinet.

5. The receptacle must be supplied by an individual branch circuit. Although the Code is silent on the issue, it would be wise to provide a 20-ampere branch circuit to accommodate any size appliance. *Individual branch circuit* is defined in Article 100 as "a branch circuit that supplies only one utilization equipment."

All serviceable components can be repaired or serviced with the hood fan in place.

422.31(B)

Disconnection, Appliances Rated Over 300-Volt-Amperes or $^1/_8$ Horsepower
NEC page 267

CHANGE TYPE: Revision (Proposal 17-24; Comment 17-83)

CHANGE SUMMARY: The revision to this section deletes the previous permission to install a temporary locking attachment on or at a circuit breaker. The same requirement was added to the 430.102(B) Exception for disconnecting means for motors and motor controllers in the 2002 NEC cycle.

2005 CODE: "**(B) Appliances Rated Over 300-Volt-Amperes or $^1/_8$ Horsepower.** For permanently connected appliances rated over 300 volt-amperes or $^1/_8$ hp, the branch-circuit switch or circuit breaker shall be permitted to serve as the disconnecting means where the switch or circuit breaker is within sight from the appliance or is capable of being locked in the open position. The provision for locking or adding a lock to the disconnecting means shall be installed on or at the switch or circuit breaker used as the disconnecting means and shall remain in place with or without a lock installed."

CHANGE SIGNIFICANCE: The problem with the previous wording of this section was that the disconnect in many appliance applications is a circuit breaker in a panelboard or a switch that is not made with permanent provisions for locking the circuit breaker or switch in the open position. This clearly does not meet the requirements of 422.31(B), which states that the switch or circuit breaker must be "capable of being locked in the open position." With lock in hand, an installer/maintainer should be able to apply it and work safely.

Previous language required a service person to carry dozens of different accessory devices to safely lock out motor power sources. This is not practical. Permanent provisions for making circuit breakers and switches capable of being locked in the open position are readily

FIGURE 4.18

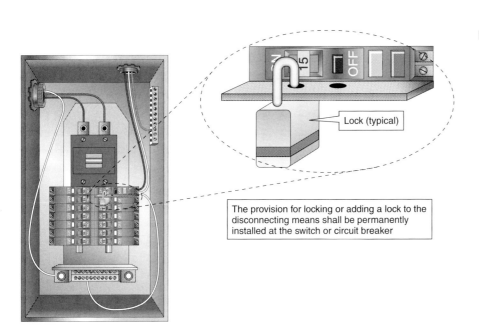

Lock (typical)

The provision for locking or adding a lock to the disconnecting means shall be permanently installed at the switch or circuit breaker

available from manufacturers today. Therefore, this proposal does not represent a large increase in the cost of an installation but will result in a dramatic increase in safety. The Code must ensure only that a lock is needed by an installer/maintainer to work safely.

The Panel accepted the same change to this section regarding the nature of the locking mechanism as was accepted for the 430.102(B) Exception. It mandates, "The provision for locking or adding a lock to the disconnecting means shall be installed on or at the switch or circuit breaker used as the disconnecting means and shall remain in place with or without a lock installed place"—at all times. This language essentially eliminates the portable lock-off accessory that a service person might otherwise carry from one location to another. These temporary devices are usually plastic and placed on the circuit breaker after placing the breaker in the off position. The placement of the lock *on* the accessory is intended to make it difficult or impossible to remove it while locked off.

Acceptable under the revised rules are lock-off slots on switches or circuit breakers and accessories supplied by the manufacturer that are installed on the breaker and remain in place all the time. Some of these accessories are made of sheet metal, snap onto the circuit breaker, and are held in place by the dead front or panelboard cover. Note that the language no longer requires the locking mechanism to be permanently installed on only the breaker. These locking provisions are usually furnished as a normal part of an enclosed safety switch.

CHANGE TYPE: New (Proposal 17-6; Comment 17-3)

CHANGE SUMMARY: Cord-and-plug-connected vending machines manufactured or re-manufactured after January 1, 2005, are required to be provided with or protected by a ground-fault circuit interruption (GFCI) device.

2005 CODE: "**422.51   Cord-and-Plug-Connected   Vending Machines**. Cord-and-plug-connected vending machines manufactured or re-manufactured on or after January 1, 2005, shall include a ground-fault circuit-interrupter as an integral part of the attachment plug or located in the power supply cord within 300 mm (12 in.) of the attachment plug. Cord-and-plug-connected vending machines not incorporating integral GFCI protection shall be connected to a GFCI protected outlet."

CHANGE SIGNIFICANCE: The U.S. Consumer Product Safety Commission (CPSC) proposed that cord-and-plug-connected vending machines be protected by a GFCI device based upon their investigation of four electrocutions in four separate incidents, three of which occurred since 1995. Two of the deaths involved children, ages 9 and 10, who were electrocuted when they contacted the vending machine.

CPSC also investigated three additional incidents with vending machines, cases that involved non-fatal, electric shocks. In all incidents, a vending machine conductor intended to carry current apparently faulted to the exposed frame of the machine, and the ground-fault path was damaged or inadequate.

CPSC stated that electric vending machines are often located in damp and wet locations, in public places, and used by people standing

422.51
Cord-and-Plug-Connected Vending Machines
NEC page 269

FIGURE 4.19

on the ground. Under these circumstances, reliance on equipment grounding conductors alone for protection against electrocution is insufficient.

Some incidents of shock and electrocution are the result of product modification that defeated the grounding feature. However, a ground-fault circuit interrupter (GFCI) does not rely on the presence of a grounding conductor to provide electrocution protection.

Given the life expectancy of the machines (10 or more years), and the likelihood that existing machines will be reconditioned or re-manufactured to extend their life, the proposal includes providing electrocution protection for machines built prior to incorporating electrocution protection as part of the machine itself. In accordance with the proposal, older machines would be connected to receptacle outlets provided with GFCI protection.

CHANGE TYPE: Revision (Proposal 17-30)

CHANGE SUMMARY: Fixed electric space-heating equipment is now required to be considered a continuous load. This section previously implied that the equipment load is continuous but was not explicit.

2005 CODE: "**(B) Branch-Circuit Sizing.** <u>Fixed electric space-heating equipment shall be considered continuous load.</u> ~~The ampacity of the branch-circuit conductors and the rating or setting of overcurrent protective devices supplying fixed electric space-heating equipment consisting of resistance elements with or without a motor shall not be less than 125 percent of the total load of the motors and the heaters. The rating or setting of overcurrent protective devices shall be permitted in accordance with 240.4(B). A contactor, thermostat, relay, or similar device, listed for continuous operation at 100 percent of its rating, shall be permitted to supply its full-rated load as provided in 210.19(A), Exception.~~

~~The size of the branch-circuit conductors and overcurrent protective devices supplying fixed electric space-heating equipment, including a hermetic refrigerant motor compressor with or without resistance units, shall be computed in accordance with 440.34 and 440.35. The provisions of this section shall not apply to conductors that form an integral part of approved fixed electric space-heating equipment.~~"

CHANGE SIGNIFICANCE: The substantiation for this proposal maintained that the language in the section has long suggested that branch-circuit overcurrent protection and conductors should be a continuous load but has stopped short of stating that. As can be seen above, the Code Panel deleted all previous language and replaced it with the simple statement that electric space-heating equipment is considered to be a continuous load.

424.3(B)

Fixed Electric Space-Heating Equipment, Branch-Circuit Sizing

NEC page 270

FIGURE 4.20

Section 210.19(A)(1) requires that the branch-circuit conductors for a continuous load be rated not less than 125 percent of the load. If a combination load is served, the conductors are required to be sized at 100 percent of the non-continuous load and 125 percent of the continuous load. Section 210.19(A)(1) also requires that overcurrent protection for the branch circuit be rated not less than 100 percent of the non-continuous load and 125 percent of the continuous load so that the requirements are coordinated.

For feeders, 215.2(A)(1) contains similar language to Article 210. Section 215.2(A)(1) requires that the minimum feeder size, before the application of any adjustment or correction factors, shall have an allowable ampacity not less than the non-continuous load plus 125 percent of the continuous load.

Similar rules on overcurrent protection of feeders can be found in 215.3. Section 220.51 contains specific rules on calculating fixed electric space-heating loads and provides that unless demand factors are granted by the authority having jurisdiction, the heating loads must be calculated at 100 percent of the total connected load. The optional calculations for dwelling units in 220.82 allows a demand factor to be applied to the fixed electric space-heating loads under some conditions.

430.6(A)(1), Tables 430.248, 430.249, 430.250

Ampacity and Motor Rating Determinations, General Motor Applications, Table Values

NEC pages 285, 310, 311

CHANGE TYPE: Revision (Proposal 11-88a; Comment 11-55)

CHANGE SUMMARY: Proposal 11-88a was written by CMP-11 in response to Proposal 11-15, which was submitted by the Technical Correlating Committee relative to Comment 11-6 and held over from the 2002 cycle. Proposal 11-88a, affecting Tables 430.148, 430.149, 430.150, and Section 430.6(A)(1), was developed to require the use of the nameplate current rating for low speed, high torque motors instead of the table values. This requirement was relocated from the table headers and placed in Section 430.6(A)(1).

2005 CODE: "**(A) General Motor Applications.** For general motor applications, current ratings shall be determined based on (A)(1) and (A)(2).

(1) Table Values. <u>Other than for motors built for low speeds (less than 1200 RPM) or high torques, and for multispeed motors,</u> the values given in Table <u>430.247</u> 430.147, Table <u>430.248</u> 430.148, Table <u>430.249</u> 430.149, and Table <u>430.250</u> 430.150 including notes shall be used to determine the ampacity of conductors or ampere ratings of switches, branch-circuit short-circuit, and ground-fault protection, instead of the actual current rating marked on the motor nameplate. Where a motor is marked in amperes, but not horsepower, the horsepower rating shall be assumed to be that corresponding to the value given in Table <u>430.247</u> 430.147, Table <u>430.248</u> 430.148, Table <u>430.249</u> 430.149, and Table <u>430.250</u> 430.150, interpolated if necessary. <u>Motors built for low speeds (less than 1200 RPM) or high torques may have higher full-load currents, and multispeed motors will have full-load currents varying with speed, in which case the nameplate current ratings shall be used.</u>"

(No change to Exceptions Nos. 1, 2, or 3.)

FIGURE 4.21

"**Table 430.248 Full-Load Currents in Amperes, Single-Phase Alternating-Current Motors.** The following values of full-load currents are for motors running at usual speeds and motors with normal torque characteristics. ~~Motors built for especially low speeds or high torques may have higher full-load currents, and multispeed motors will have full-load current varying with speed, in which case the nameplate current ratings shall be used.~~

~~The voltages listed are rated motor voltages.~~ The currents listed shall be permitted for system voltage ranges of 110 to 120 and 220 to 240 volts.

Table 430.249 Full-Load Current, Two-Phase Alternating-Current Motors (4-Wire). The following values of full-load current are for motors running at speeds usual for belted motors and motors with normal torque characteristics. ~~Motors built for especially low speeds or high torques may require more running current, and multispeed motors will have full-load current varying with speed, in which case the nameplate current rating shall be used.~~ Current in the common conductor of a 2-phase, 3-wire system will be 1.41 times the value given.

The voltages listed are rated motor voltages. The currents listed shall be permitted for system voltage ranges of 110 to 120, 220 to 240, 440 to 480, and 550 to 600 volts.

Table 430.250 Full-Load Current, Three-Phase Alternating-Current Motors.

The following values of full-load currents are typical for motors running at speeds usual for belted motors and motors with normal torque characteristics.

~~Motors built for low speeds (1200 rpm or less) or high torques may require more running current, and multispeed motors will have full-load current varying with speed. In these cases, the nameplate current rating shall be used.~~

The voltages listed are rated motor voltages. The currents listed shall be permitted for system voltage ranges of 110 to 120, 220 to 240, 440 to 480, and 550 to 600 volts."

CHANGE SIGNIFICANCE: The Code Panel's substantiation for the proposal indicates that deleting the sentence beginning "Motors built for ..." from the Table Headers in 430.148, 149, and 150 and adding the same sentence as a new last sentence to 430.6(A)(1) adds clarity. Deleting the words "including notes" from the first sentence of 430.6(A)(1) further adds clarity to this section.

430.8, Exceptions

Marking on Controllers
NEC page 287

CHANGE TYPE: Revision and New (Proposal 11-19; Comment 11-16)

CHANGE SUMMARY: Motor controllers are now required to be marked with the "short-circuit current rating." New exceptions are provided for certain motor controller applications.

2005 CODE: "**430.8 Marking on Controllers.** A controller shall be marked with the manufacturer's name or identification, the voltage, the current or horsepower rating, <u>the short-circuit current rating,</u> and such other necessary data to properly indicate the <u>applications</u> ~~motors~~ for which it is suitable.

<u>Exception No. 1: The short-circuit current rating is not required for controllers applied in accordance with 430.81(A), or 430.83(B).</u>

<u>Exception No. 2: The short-circuit current rating is not required to be marked on the controller when the short-circuit current rating of the controller is marked elsewhere on the assembly.</u>

<u>Exception No. 3: The short-circuit current rating is not required to be marked on the controller when the assembly into which it is installed has a marked short-circuit current rating.</u>

<u>Exception No. 4: Short-circuit ratings are not required for controllers rated less than 2 hp at 300V or less and listed exclusively for general-purpose branch-circuits.</u>

A controller that includes motor overload protection suitable for group motor application shall be marked with the motor overload protection and the maximum branch-circuit short-circuit and ground-fault protection for such applications.

Combination controllers that employ adjustable instantaneous trip circuit breakers shall be clearly marked to indicate the ampere settings of the adjustable trip element.

Where a controller is built-in as an integral part of a motor or of a motor-generator set, individual marking of the controller shall not be

FIGURE 4.22

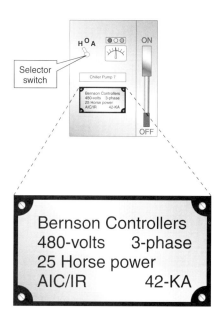

Bernson Controllers
480-volts   3-phase
25 Horse power
AIC/IR   42-KA

required if the necessary data are on the nameplate. For controllers that are an integral part of equipment approved as a unit, the above marking shall be permitted on the equipment nameplate.

FPN: See 110.10 for information on circuit impedance and other characteristics."

CHANGE SIGNIFICANCE: The substantiation for this proposal indicates that controllers are subjected to significant amounts of available short-circuit current and must therefore have an adequate short-circuit current rating for these faults. Fortunately, UL 508, Industrial Control Equipment, provides a means to mark this short-circuit current rating.

The phrase "and such other necessary data to properly indicate the applications for which it is suitable" is vague, even with the other changes made. The intention is that a person will be able to determine the type of motor the controller is suitable for, not whether the controller is suitable for a motor used on a sausage grinder, for example. Other common markings on the controller include the Hz (cycles) and a table for selection of heaters for the overload relays.

Exception No. 1 was added to exempt certain controllers for small motors as allowed in Part VII, such as clock motors, attachment plugs and receptacles, and snap switches. The second exception was added to permit controller short-circuit current ratings to be marked elsewhere on the assembly. The third exception was added because a short-circuit current rating on an assembly covers all of the internal components.

430.32(A)

Continuous-Duty Motors, More Than 1 Horsepower
NEC page 292

CHANGE TYPE: Revision (Proposal 11-30)

CHANGE SUMMARY: The changes to this section indicate that motors used in a continuous-duty application require overload protection. Continuous-duty motors used in non-continuous applications do not require the same protection.

2005 CODE: "**(A) More Than 1 Horsepower.** Each motor <u>used in a continuous-duty application and</u> rated more than 1 hp shall be protected against overload by one of the means in 430.32(A)(1) through (A)(4)."

CHANGE SIGNIFICANCE: The previous wording may have led individuals to think that motors are rated as "continuous-duty." The "duty" of a motor is determined by the motor's application and defined in Article 100 under *duty*. A motor rated by the manufacturer as "continuous" and used in other than continuous duty requires no overloads.

FIGURE 4.23

430.102 (B), Exception

Disconnecting Means, Motor

NEC page 304

CHANGE TYPE: Revision (Proposal 11-67; Comment 11-42)

CHANGE SUMMARY: More changes have been made to the rule on disconnecting means within sight of the motor and driven machinery. Also, the second sentence of (B) has been deleted, as well as a phrase in the exception.

2005 CODE: "**(B) Motor.** A disconnecting means shall be located in sight from the motor location and the driven machinery location. ~~The disconnecting means required in accordance with 430.102(A) shall be permitted to serve as the disconnecting means for the motor if it is located in sight from the motor location and the driven machinery location.~~

Exception: The disconnecting means shall not be required to be in sight from the motor and the driven machinery location under either condition (a) or (b), provided the disconnecting means required in accordance with 430.102(A) is individually capable of being locked in the open position. The provision for locking or adding a lock to the disconnecting means shall be ~~permanently~~ installed on or at the switch or circuit breaker used as the disconnecting means <u>and shall remain in place with or without the lock installed</u>.

(a) Where such a location of the disconnecting means is impracticable or introduces additional or increased hazards to persons or property

(b) In industrial installations, with written safety procedures, where conditions of maintenance and supervision ensure that only qualified persons service the equipment

FPN NO. 1:   Some examples of increased or additional hazards include, but are not limited to, motors rated in excess of 100 hp, multimotor equipment, submersible motors, motors associated with <u>adjustable</u>

FIGURE 4.24

speed drivers ~~variable frequency drives~~, and motors located in hazardous (classified) locations.

FPN NO. 2:   For information on lockout/tagout procedures, see NFPA 70E-2004, Standard for Electrical Safety in the Workplace.

The disconnecting means required in accordance with 430.102(A) shall be permitted to serve as the disconnecting means for the motor if it is located in sight from the motor location and the driven machinery location."

CHANGE SIGNIFICANCE: The Code Panel has made organizational and substantive changes to the section's main text and exception. The second sentence of the opening paragraph has been moved to follow the Fine Print Notes so that it is no longer modified by the exception. Now, clearly, the disconnecting means on the supply side and within sight of the motor controller is permitted to be the only disconnecting means for the controller and motor if it is within sight of both.

The word "permanently" in the exception, as related to the means for locking the disconnecting means is concerned, has been deleted as it was considered confusing. The intent of this change in the 2002 NEC was to require the provisions for locking to always be in place, even when the lock is removed. "Permanently" was used to convey this; however, "permanently" has also been interpreted to mean that it shall not be possible to remove the locking means. ("That was not the intent and there are few, if any, locking means that are permanent to the point they cannot be removed.")

The phrase "and shall remain in place with or without the lock installed" effectively removes any permission to use an accessory that can be moved from disconnect to disconnect. These devices are usually plastic and are placed on the circuit breaker after placing the breaker in the off position. The placement of the lock *on* the accessory is intended to make it difficult or impossible to remove it while locked off. Now a lock is all that is necessary for a worker to lock-out the disconnecting means and work on the equipment safely.

430.109 (A)(6)

Manual Motor Controller

NEC page 305

CHANGE TYPE: Revision (Proposal 11-73; Comment 11-45)

CHANGE SUMMARY: Manual motor controllers are now permitted as the motor disconnecting means where so marked. Rules have been added on locating the manual motor starter on the line side of the fuses used for running overload protection as permitted in 430.52(C)(5).

2005 CODE: "**(6) Manual Motor Controller.** Listed manual motor controllers additionally marked "Suitable as Motor Disconnect" shall be permitted as a disconnecting means where installed between the final motor branch-circuit short-circuit protective device and the motor. Listed manual motor controllers additionally marked "Suitable as Motor Disconnect" shall be permitted as disconnecting means on the line side of the fuses permitted in 430.52(C)(5). In this case, the fuses permitted in 430.52(C)(5) shall be considered supplementary fuses, and suitable branch-circuit short-circuit and ground-fault protective devices shall be installed on the line side of the manual motor controller additionally marked "Suitable as Motor Disconnect.""

CHANGE SIGNIFICANCE: Semiconductor fuses are permitted to be used as branch-circuit fuses under 430.52(C)(5). However, they are often used as supplementary fuses, to protect electronic equipment, and may be located on the load side of listed manual motor controllers marked as "Suitable for Motor Disconnect." Since, as branch-circuit fuses, they are technically the final motor branch-circuit protective device, use past these manual motor controllers violates the Code. The proposed change permits the use of these fuses in this location, as supplementary protective devices. The manual motor controllers additionally marked "Suitable as Motor Disconnect" will be suitably protected by the branch-circuit protective devices located on their line side.

FIGURE 4.25

CHANGE TYPE: New (Proposal 11-74; Comment 11-45a)

CHANGE SUMMARY: A definition of the term *system isolation equipment* has been added to 430.2. A new subsection has been added to 430.109(A) to control the installation of the system.

2005 CODE: "**430.2 System Isolation Equipment.** A redundantly monitored, remotely operated contactor-isolating system, packaged to provide the disconnection/isolation function, capable of verifiable operation from multiple remote locations by means of lockout switches, each having the capability of being padlocked in the "off" (open) position."

"**430.109(A)(7) System Isolation Equipment.** System isolation equipment shall be listed for disconnection purposes. System isolation equipment shall be installed on the load side of the overcurrent protection and its disconnecting means. The disconnecting means shall be one of the types permitted by 430.109(A)(1) through (A)(3)."

CHANGE SIGNIFICANCE: This proposal was intended to align with the latest edition of NFPA 79, specifically 5.5 Devices for Disconnecting (Isolating) Electrical Equipment, Paragraph 5.54 (3). The described equipment is principally intended for industrial machines, covered by NFPA 79, where, because of multiple entry points or high frequency of usage, the use of other isolation devices becomes impracticable.

Because of size, manufacturing machines often have several entry points used by operators and maintenance personnel who cannot always see each other. With a "monitored safety lockout system" each point-of-entry has a "lockout capable disconnecting means," and a method to verify to the user that the disconnection function has succeeded. Each of the several point-of-entry "disconnecting means" is monitored and opens a magnetic contactor that prevents the mass production industrial machine from being energized.

430.2 and 430.109 (A)(7)

System Isolation Equipment
NEC pages 285, 305

FIGURE 4.26

The system (equipment), including the controls and contactor(s), is physically and (with the exception of the power circuit that is being controlled) electrically isolated from the rest of the industrial machine. This isolation feature is intended to ensure the safety of the operators and maintenance personnel using the isolation system (equipment). Although referred to as a "system," the equipment consists of a collection of components contained as a single rated piece of equipment that includes the contactor(s) as the machine power isolating component, the monitoring component(s), the multiple local lockout-capable disconnecting-means components, and the component(s) used to verify the disconnection function.

The magnetic contactor of such a system (equipment) cannot be again closed until all the point-of-entry disconnecting means are closed. Therefore, the machine cannot be energized accidentally with someone still working on it. The use of a magnetic contactor that isolates the industrial machine's power lends itself to this application because of the relatively long electrical and mechanical life of contactors nominally available in today's market.

Although sized to be capable of operating under full load and allow coordination of overcurrent protection devices, the "redundantly monitored, remotely operated contactor-isolating system" would normally be operated when the machine is in a stopped mode. The function and performance of a contactor can be monitored if performance (isolation) verification indications are given to the operators and maintenance personnel use the isolation system (equipment) according to these recommendations.

Presently, other disconnecting means and manual controllers are often not suitable for applications requiring numerous opening and closing operations, nor for applications requiring the multiple operating points that are often required on many industrial machines used in manufacturing. A failure of the traditional disconnecting means usually means a tremendous loss of production and possible undetected, unsafe conditions. In many situations, it is difficult to provide the machine operators and maintenance personnel with a simple way to verify that the disconnecting means has not failed; this is especially difficult when there is more than one point of entry being used during the course of industrial machine maintenance activity. The "redundantly monitored, remotely operated contactor-isolating system" is always located within the machine's electrical system on the load side of a circuit breaker or fuse.

Typically, a "redundantly monitored, remotely operated contactor-isolating system" that incorporates control lockout provisions has been reviewed with OSHA and interpreted as being a suitable part of an energy control program, and therein a suitable disconnecting means for such purposes.

Additional analysis of typical "redundantly monitored, remotely operated contactor-isolating systems" as a component of an energy control program has also been done in other independent studies. Analysis of typical forms of the proposed technology has shown a sig-

nificant advantage of reduced risk of injury when compared to existing technologies.

As the need becomes clearer, Underwriters Laboratories is expected
to begin the process of developing one or more standards for a "redundantly monitored, remotely operated contactor-isolating system" that
incorporates control lockout provisions and provides a disconnecting
means technology.

430.2 and Article 430 Part X

Adjustable Speed Drives

NEC pages 284, 307

CHANGE TYPE: New (Proposals 11-6 and 11-10; Comments 11-50 and 11-51)

CHANGE SUMMARY: New definitions have been added to 430.2 for *adjustable speed drive* and *adjustable-speed drive system*. Additionally, a new Part X has been added to Article 430 to address the issues of adjustable-speed drive systems. This change places requirements for these drive systems in a central location in Article 430 (beginning at 430.120).

2005 CODE: "**430.2 Adjustable Speed Drive.** A combination of the power converter, motor, and motor-mounted auxiliary devices such as encoders, tachometers, thermal switches and detectors, air blowers, heaters, and vibration sensors."

"**430.2 Adjustable-Speed Drive System**. An interconnected combination of equipment that provides a means of adjusting the speed of a mechanical load coupled to a motor. A drive system typically consists of an adjustable speed drive and auxiliary electrical apparatus."

"X. Adjustable-Speed Drive Systems
430.120 General
430.122 Conductors–Minimum Size and Ampacity
(A) Branch/Feeder Circuit Conductors
(B) Bypass Device
430.124 Overload Protection
(A) Included in Power Conversion Equipment
(B) Bypass Circuits

FIGURE 4.27

(C) Multiple Motor Applications
430.126 Motor Overtemperature Protection
(A) General
(B) Motors with Cooling Systems
(C) Multiple Motor Applications
(D) Automatic Restarting and Orderly Shutdown
430.128 Disconnecting Means."

CHANGE SIGNIFICANCE: The substantiation for this proposal stated that adjustable-speed drive systems have gained enormous popularity over the last few years, and so appropriate installation rules need to be added to cover these systems. A companion proposal was accepted to add definitions of two terms pertaining to drive systems to 430.2. In response to the initial proposal, CMP-11 established a specific part of Article 430 to address the key rules for installation of adjustable-speed drive systems. Companion proposals were submitted to move material pertaining to drives from 430.2 and 430.22 Exception No. 2 to this new Part X.

As a result of this new part of Article 430, previous parts were renumbered, as were several sections and tables that follow new Part X. Part XI now covers "Over 600 Volts, Nominal" and begins with 430.221. Part XII now covers "Protection of Live Parts—All Voltages" and begins with 430.231. Part XIII now covers Grounding—All Voltages" and begins with 430.241. Part XIV now covers tables and begins with Table 430.247 rather than Table 430.147.

440.4(B), Exception

Multimotor and Combination-Load Equipment

NEC page 314

CHANGE TYPE: Revision and New (Proposal 11-95; Comment 11-59)

CHANGE SUMMARY: This proposal adds a requirement for marking short-circuit current rating, as well as a related exception.

2005 CODE: "**(B) Multimotor and Combination-Load Equipment.** Multimotor and combination-load equipment shall be provided with a visible nameplate marked with the maker's name; the rating in volts, frequency, and number of phases, minimum supply circuit conductor ampacity, ~~and~~ the maximum rating of the branch-circuit short-circuit and ground-fault protective device, <u>and the short-circuit current rating of the motor controllers or industrial control panel</u>. The ampacity shall be calculated by using Part IV and counting all the motors and other loads that will be operated at the same time. The branch-circuit short-circuit and ground-fault protective device rating shall not exceed the value calculated by using Part III. Multimotor or combination-load equipment for use on two or more circuits shall be marked with the above information for each circuit.

Exception No. 1: Multimotor and combination-load equipment that is suitable under the provisions of this article for connection to a single 15- or 20-ampere, 120-volt, or a 15-ampere, 208- or 240-volt, single-phase branch circuit shall be permitted to be marked as a single load.

Exception No. 2: The minimum supply circuit conductor ampacity and the maximum rating of the branch-circuit short-circuit and ground-fault protective device shall not be required to be marked on a room air conditioner conforming with 440.62(A).

<u>*Exception No. 3: Multimotor and combination-load equipment used in one- and two-family dwellings, cord-and-attachment-plug connected equipment, or equipment supplied from a branch circuit protected at 60 A or less shall not be required to be marked with a short-circuit current rating.*</u>"

FIGURE 4.28

CHANGE SIGNIFICANCE: The substantiation for the proposal states that HVAC equipment installed in commercial, industrial, and multifamily dwellings may be subject to significant amounts of available short-circuit current. Not only must the overcurrent protective devices(s) have an adequate interrupting rating, but the other electrical components must also have an adequate short-circuit current rating. Therefore, the equipment must be marked with the amount of short-circuit current acceptable for use. Without this marking, the installer is unable to determine if the equipment being supplied is suitable for the application. Fortunately, the new UL 508A, Industrial Control Panels, now provides a means to mark the short-circuit current rating on HVAC equipment.

Opponents of the marking requirement have maintained that the concept of containment alone provides the level of safety required, that the installer does not need to know the short-circuit current rating of the end product for a safe installation. In the event a short circuit occurs in the listed end product and the short-circuit protection provided by the installer does not open before component failure occurs, the enclosure will contain the results of the failure. The end product may be damaged, but the shock and fire hazard will be contained within the end-product enclosure.

This is a worker safety issue. It is important that the short-circuit current rating of the equipment be known so that the equipment can be installed safely. As previously noted, motor controllers are now required to be marked with their short-circuit current rating. New Article 409 also requires the equipment to be marked for its short-circuit current rating. Since motor controllers and industrial control panels are required to be marked, requiring a similar marking for control equipment and HACR equipment is the next logical step.

As indicated, equipment for one- and two-family dwellings as well as equipment with a branch circuit rated 60 amperes or less are not required to be marked with a short-circuit current rating—this equipment would usually have a limited amount of short-circuit current available.

440.14 Exception No. 1

Disconnecting Means, Location

NEC page 316

CHANGE TYPE: Revision (Proposal 11-100)

CHANGE SUMMARY: Changes to this section include requiring an industrial facility that hopes to comply with Exception No. 1 to have written safety procedures. In addition, the provisions for locking or adding a lock to the disconnecting means shall now be permanently installed on or at the switch or circuit breaker used as the disconnecting means.

2005 CODE: "**440.14 Location.** Disconnecting means shall be located within sight from and readily accessible from the air-conditioning or refrigerating equipment. The disconnecting means shall be permitted to be installed on or within the air-conditioning or refrigerating equipment.

The disconnecting means shall not be located on panels that are designed to allow access to the air-conditioning or refrigeration equipment.

Exception No. 1: Where the disconnecting means provided in accordance with 430.102(A) is capable of being locked in the open position, and the refrigerating or air-conditioning equipment is essential to an industrial process in a facility <u>with written safety procedures, and </u>where the conditions of maintenance and the supervision ensure that only qualified persons service the equipment, a disconnecting means within sight from the equipment shall not be required. <u>The provisions for locking or adding a lock to the disconnecting means shall be permanently installed on or at the switch or circuit breaker used as the disconnecting means.</u>

Exception No. 2: Where an attachment plug and receptacle serve as the disconnecting means in accordance with 440.13, their location shall be accessible but shall not be required to be readily accessible.

FIGURE 4.29

FPN: See Parts VII and IX of Article 430 for additional requirements.

CHANGE SIGNIFICANCE: This proposal intended to apply the requirement put in place during the last Code cycle for the application of 430.102(B) Exception regarding the provision for adding a lock to the disconnecting means for motors. The need for this additional sentence is based on the fact that on some panelboards or distribution centers the locking means is rendered ineffective when the dead front or cover is removed. This can cause an unsafe condition for persons working on the equipment, which certainly contradicts the intent of the original exception. And it really makes little difference whether the disconnecting means is for a motor or for a component of electric motor-driven air-conditioning and refrigerating equipment—the requirement should apply because the same hazard exists.

The addition of a permanent means on or at the switch or circuit breaker ensures that the lockout can be successfully performed without the fear that the lock will be inadvertently removed. It should be noted that the Occupational Safety and Health Administration (OSHA) is presently citing employers for the use of many of the plastic, non-listed, snap-on circuit breaker and switch lockout devices.

The proposal also includes another requirement. In order to apply Exception No. 1 and avoid being required to have a disconnecting means within sight of the air-conditioning or refrigerating equipment, the industrial facility is required to have written safety procedures.

440.32

Single Motor Compressor

NEC page 317

CHANGE TYPE: Revision (Proposal 11-105)

CHANGE SUMMARY: These changes mandate that the previous 58 percent permitted for wye-start, delta-run motors is to be multiplied by 125 percent.

2005 CODE: "**440.32 Single Motor-Compressor.** Branch-circuit conductors supplying a single motor-compressor shall have an ampacity not less than 125 percent of either the motor-compressor rated-load current or the branch-circuit selection current, whichever is greater.

For a wye-start, delta-run connected motor-compressor, the selection of branch-circuit conductors between the controller and the motor-compressor shall be permitted to be based on 72 58 percent of either the motor-compressor rated-load current or the branch-circuit selection current, whichever is greater."

FPN: The individual motor circuit conductors of wye-start, delta-run connected motor-compressors carry 58 percent of the rated load current. The multiplier of 72 percent is obtained by multiplying 58 percent by 1.25.

CHANGE SIGNIFICANCE: This change clears up the debate on how to properly size the conductors from a six-lead motor to the controller. whether the rules in 430.22(A) and 440.32, which generally require a 125 percent factor, should be applied to motor branch-circuit conductors. As indicated, the Code Panel has applied the 125 percent factor to the 58 percent load current to reach the 78 percent rating.

FIGURE 4.30

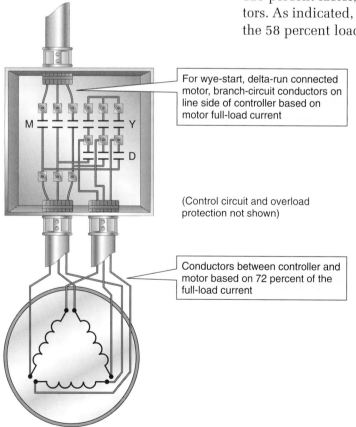

For wye-start, delta-run connected motor, branch-circuit conductors on line side of controller based on motor full-load current

M   Y

D

(Control circuit and overload protection not shown)

Conductors between controller and motor based on 72 percent of the full-load current

CHANGE TYPE: Revision (Proposal 13-5)

CHANGE SUMMARY: The items required to be identified on generator nameplates now include the subtransient and transient impedances of the generator.

2005 CODE: "**445.11 Marking**. Each generator shall be provided with a nameplate giving the manufacturer's name, the rated frequency, power factor, number of phases if of alternating current, <u>the subtransient and transient impedances,</u> the rating in kilowatts or kilovolt amperes, the normal volts and amperes corresponding to the rating, rated revolutions per minute, insulation system class and rated ambient temperature or rated temperature rise, and time rating."

CHANGE SIGNIFICANCE: This change was proposed to provide additional information required to calculate the short-circuit current available from generators. Revisions to NFPA 70E will reportedly require calculation of arc energy for electrical equipment. The generator impedance value will be needed for this calculation.

445.11
Marking
NEC page 320

FIGURE 4.31

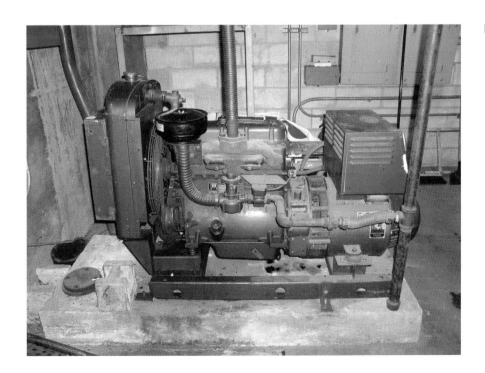

445.18

Disconnecting Means Required for Generators

NEC page 320

CHANGE TYPE: Revision (Proposal 13-7)

CHANGE SUMMARY: Revisions to this section on disconnecting means now permit more than one form of disconnect so that the generator can supply different services or systems at the same time. Multiple disconnecting means also allow the systems supplied to be isolated individually without being disconnected at the same time.

2005 CODE: "**445.18 Disconnecting Means Required for Generators.** Generators shall be equipped with a̶ disconnect(s) by means of which the generator and all protective devices and control apparatus are able to be disconnected entirely from the circuits supplied by the generator except where both of the following conditions apply:

(1) The driving means for the generator can be readily shut down; and̶.

(2) The generator is not arranged to operate in parallel with another generator or other source of voltage.

CHANGE SIGNIFICANCE: If these provisions for additional disconnects are employed, then two or three circuit breakers at the generator feeding different transfer switches could be installed. This would add needed flexibility in certain situations. It would also allow portions of the emergency system to be more easily maintained without the need to take the entire system out of service, and, given the difficulty in performing needed shutdowns in facilities like hospitals and data centers, the overall reliability of those systems would be enhanced due to the (relative) ease in arranging such necessary shutdowns.

FIGURE 4.32

CHANGE TYPE: Revision (Proposal 9-132; Comment 9-133)

CHANGE SUMMARY: A sentence has been added to clarify that zigzag grounding autotransformers are not permitted to be installed on the load side of a grounded system.

2005 CODE: "**450.5 Grounding Autotransformers.** Grounding autotransformers covered in this section are zigzag or T-connected transformers connected to 3-phase, 3-wire ungrounded systems for the purpose of creating a 3-phase, 4-wire distribution system or providing a neutral reference for grounding purposes. Such transformers shall have a continuous per-phase current rating and a continuous neutral current rating. <u>Zigzag connected transformers shall not be installed on the load side of any system grounding connection, including those made in accordance with 250.24(B), 250.30(A)(1), or 250.32(B)(2).</u>"

CHANGE SIGNIFICANCE: The intent of this proposal was to prohibit the installation of a zigzag transformer on a 3-phase, 4-wire wye-connected system. Applying a zigzag connection on the 3-wire branch circuit of a 4-wire wye system will result in any ground fault on the system sharing fault current through the zigzag transformer. The Code Panel rejected the proposal.

The new sentence added instead by the Panel prohibits a zigzag grounding connection downstream from another system grounding connection, such as a service or a separately derived system. Though it us not considered a "system" grounding connection, the new sentence prohibits making a zigzag grounding connection on the load side of a building disconnecting means.

Zigzag grounding autotransformers are typically installed to create a grounded system from a 3-phase, 3-wire delta-connected ungrounded system.

450.5
Grounding Autotransformers
NEC page 322

FIGURE 4.33

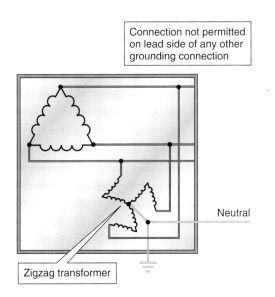

Connection not permitted on lead side of any other grounding connection

Neutral

Zigzag transformer

490.46

Metal-Enclosed and Metal-Clad Service Equipment

NEC page 337

CHANGE TYPE: New (Proposal 9-141; Comment 9-136)

CHANGE SUMMARY: A new section has been added to cover metal-enclosed and metal-clad switchgear used as service equipment for supplies that are over 600 volts.

2005 CODE: "**490.46 Metal-Enclosed and Metal-Clad Service Equipment.** Metal-enclosed and metal-clad switchgear installed as high-voltage service equipment shall include a ground bus for the connection of service cable shields and to facilitate the attachment of safety grounds for personnel protection. This bus shall be extended into the compartment where the service conductors are terminated."

FIGURE 4.34

CHANGE SIGNIFICANCE: The revision suggested initially was for a fairly comprehensive set of requirements for service equipment operating above 600 volts. The proposal was rejected on the basis that "These provisions are specific to one utility or geographic area of the country and are not representative of all utility requirements. Additionally, the majority of the requirements are normally supplied in metal-clad switchgear but are not necessarily included in metal-enclosed switchgear. Those utilities accepting metal-enclosed type equipment are not requiring all these features and may not want to bear the additional costs associated with this proposal." The new language identified above was accepted at the comment meeting.

It should be noted that Part VIII of Article 230 contains several requirements for installing services over 600 volts.

CHAPTER **5**

Special Occupancies, Articles 500–590

500.2
Purged and Pressurized
NEC page 341

CHANGE TYPE: New (Proposal 14-8; Comment 14-6)

CHANGE SUMMARY: A definition for *purged and pressurized* has been added to 500.2. Text extracted from NFPA 496 has also been added, and clarifies that the protection technique includes both purging and pressurizing the enclosure.

2005 CODE: "**Purged and Pressurized.** The process of (1) purging, supplying an enclosure with a protective gas at a sufficient flow and positive pressure to reduce the concentration of any flammable gas or vapor initially present to an acceptable level; and (2) pressurization, supplying an enclosure with a protective gas with or without continuous flow at sufficient pressure to prevent the entrance of a flammable gas or vapor, a combustible dust, or an ignitable fiber.

FPN: For further information, see ANSI/NFPA 496-2003, *Purged and Pressurized Enclosures for Electrical Equipment.*"

CHANGE SIGNIFICANCE: The previous paragraph contained only half of the requirements of NFPA 496, "Purging." To ensure safety, the requirements from "Pressurization" must also be included.

Purged and pressurized is a protective technique recognized in 500.7 for hazardous (classified) locations. However, information was not previously provided in the NEC for this protection technique. By comparison, intrinsic safety is identified in 500.7 as a protection technique and an entire related Article 504, "Intrinsically Safe Systems" is included in the Code.

The Fine Print Note refers to NFPA 496, the standard for *Purged and Pressurized Enclosures for Electrical Equipment*, where it is possible to find all of the requirements for installing purged or pressurized systems, including Type X, Type Y, and Type Z systems.

FIGURE 5.1

CHANGE TYPE: Revision (Proposals 14-13; 14-14)

CHANGE SUMMARY: The revised text clarifies that a gas detection system must include items listed in recommended practices that involve maintenance and operation requirements, in addition to installation requirements, for the protection technique to operate safely. All of these items must be documented before the protection technique is accepted. A similar change was made to 505.8(I) by Proposal 14-85.

2005 CODE: "**(K) Combustible Gas Detection System.** A combustible gas detection system shall be permitted as a means of protection in industrial establishments with restricted public access and where the conditions of maintenance and supervision ensure that only qualified persons service the installation. Gas detection equipment shall be listed for detection of the specific gas or vapor to be encountered. Where such a system is installed, equipment specified in 500.7(K)(1), (2), or (3) shall be permitted. The type of detection equipment, its listing, installation location(s), alarm and shutdown criteria, and calibration frequency shall be documented when combustible gas detectors are used as a protection technique.

FPN NO. 1:   For further information, see ANSI/ISA 12.13.01, Performance Requirements, Combustible Gas Detectors.

FPN NO. 2:   For further information, see ANSI/API RP 500, Recommended Practice for Classification of Locations for Electrical Installations at Petroleum Facilities Classified as Class I, Division 1, or Division 2.

FPN NO. 3:   For further information, see ISA-RP 12.13.02, Installation, Operation, and Maintenance of Combustible Gas Detection Instruments.

500.7(K)
Combustible Gas Detection System
NEC page 345

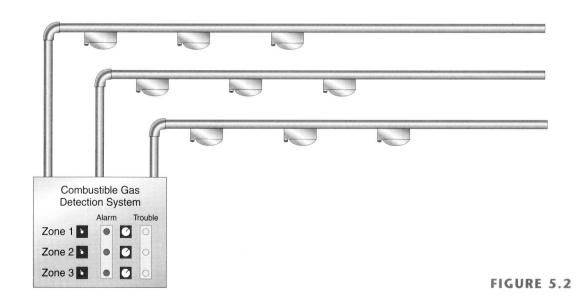

FIGURE 5.2

(1) Inadequate Ventilation. In a Class I, Division 1 location that is so classified due to inadequate ventilation, electrical equipment suitable for Class I, Division 2 locations shall be permitted.

(2) Interior of a Building. In a building located in, or with an opening into, a Class I, Division 2 location where the interior does not contain a source of flammable gas or vapor, electrical equipment for unclassified locations shall be permitted.

(3) Interior of a Control Panel. In the interior of a control panel containing instrumentation utilizing or measuring flammable liquids, gases, or vapors, electrical equipment suitable for Class I, Division 2 locations shall be permitted.

~~FPN No. 1: For further information, see ANSI/ISA 12.13.01, Performance Requirements, Combustible Gas Detectors.~~

~~FPN No. 2: For further information., see ANSI/API RP 500, Recommended Practice for Classification of Locations for Electrical Installations at Petroleum Facilities Classified as Class I, Division I or Division 2.~~

~~FPN No. 3: For further information, see ISA RP 12.13.02, Installation, Operation, and Maintenance of Combustible Gas Detection Instruments.~~"

CHANGE SIGNIFICANCE: The combustible gas detection system is recognized as a protected technique and is defined in 500.2. However, little information is provided in Article 500 for the installation of such a system. Equipment is available that is specifically designed to detect the presence of a combustible gas. It can be set to sound an alarm at a concentration usually below the level of the lower flammable limit for the gas. These systems are limited to installation in industrial establishments with restricted public access and where qualified persons service installation.

As can be seen in (K)(1), in a Class I, Division 1 location that is classified due to inadequate ventilation, electrical equipment suitable for Class 1, Division 2 is permitted to be used when a combustible gas detection system is installed. Obviously Class I, Division 2 equipment is much less expensive than Class I, Division 1 equipment.

As indicated in (K)(2), in the interior of a building usually classified as a Class I, Division 2 location where the interior does not contain a source of flammable gas or vapors, electrical equipment for an unclassified location is permitted when a combustible gas detection system is installed. Similar to Class I, Division 2 equipment, equipment for an unclassified area is obviously less expensive than the equipment suitable for a Class I, Division 2 area.

In interiors of control panels that contain instrumentation utilizing or measuring flammable liquids, gases, or vapors, electrical equipment suitable for a Class I, Division 2 location is permitted if a qualified combustible gas detection system is installed.

501.10(B) (6)

Wiring Methods, Class I, Division 2 Areas
NEC page 349

CHANGE TYPE: Revision (Proposal 14-47)

CHANGE SUMMARY: Single-conductor MV cables installed in a Class I, Division 2 area are now required to be shielded or metallic armored.

2005 CODE: "(6) Type MI, MC, MV, or TC cable with termination fittings, or in cable tray systems and installed in a manner to avoid tensile stress at the termination fittings. <u>Single conductor Type MV cables shall be shielded or metallic armored.</u>"

CHANGE SIGNIFICANCE: Note that with the renumbering that took place in Article 501, Part II (Wiring) now begins with 501.10. Requirements on wiring methods in Class I areas were found in 501.4 in the 2002 NEC.

The Type MV cable designation covers single and multiconductor constructions as well as non-shielded and shielded constructions. The proposer maintained that the use of single-conductor non-shielded MV cable should not be allowed in a Division 2 location, since the non-shielded cable will have a surface discharge from the cable surface to any ground plane (such as metal cable tray). This discharge is an ignition source that can cause an explosion in the event that gas or vapors are present in a concentration that is within its flammable limits.

The accepted wording requires single-conductor Type MV cable to have a shield or metallic armor to provide a ground plane. This ground plane will eliminate any external electrical discharge, thus eliminating the ignition source and precluding any possibility of an explosion.

FIGURE 5.3

501.15(B)(2), and Exception No. 2

Conduit Seals, Class I, Division 2 Boundary

NEC page 351

CHANGE TYPE: Revision (Proposal 14-34; Comment 14-43)

CHANGE SUMMARY: Where a boundary seal is not also completing an explosion-proof enclosure, a technical reason for the seal to be explosion-proof does not exist. In a Division 2 location, the conduit would only have gases or vapors in an abnormal condition. Therefore, the likelihood of having gases or vapors in a Division 2 location conduit and an electrical fault to provide an ignition source simultaneously is very remote. The NEC will now consider this an acceptable risk. The only purpose for the seal is to minimize gas or vapor within the Division 2 location that can pass to an unclassified area.

2005 CODE: "**501.15(B)(2) Class I, Division 2 Boundary.** In each conduit run passing from a Class I, Division 2 location into an unclassified location. The sealing fitting shall be permitted on either side of the boundary of such location within 3.05 m (10 ft) of the boundary. Rigid metal conduit or threaded steel intermediate metal conduit shall be used between the sealing fitting and the point at which the conduit leaves the Division 2 location, and a threaded connection shall be used at the sealing fitting. Except for listed reducers at the conduit seal, there shall be no union, coupling, box, or fitting between the conduit seal and the point at which the conduit leaves the Division 2 location. Conduits shall be sealed to minimize the amount of gas or vapor within the Division 2 portion of the conduit from being communicated to the conduit beyond the seal. Such seals shall not be required to be explosion-proof but shall be identified for the purpose of minimizing passage of gases under normal operating conditions and shall be accessible.

Exception No. 1: Metal conduit that contains no unions, couplings, boxes, or fittings and passes completely through a Class I, Division 2 location with no fittings less than 300 mm (12 in.) beyond each boundary, shall not be required to be sealed if the termination points of the unbroken conduit are in unclassified locations.

FIGURE 5.4

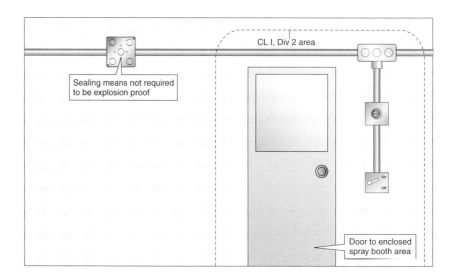

Exception No. 2: Conduit systems terminating at an unclassified location where a wiring method transition is made to cable tray, cable-bus, ventilated busway, Type MI cable, or <u>cable not installed in any cable tray or raceway system</u> ~~open wiring~~ shall not be required to be sealed where passing from the Class I, Division 2 location into the unclassified location. The unclassified location shall be outdoors or, if the conduit system is all in one room, it shall be permitted to be indoors. The conduits shall not terminate at an enclosure containing an ignition source in normal operation."

CHANGE SIGNIFICANCE: This section was previously numbered 501.5(B)(2) before the renumbering of Part II of Article 501.

Clearly, an explosion-proof seal is not required at the boundary where the conduit passes from a Class I, Division 2 area to an unclassified area. However, the seal is required to be "identified for the purpose of minimizing passage of gases under normal operating conditions." This language may in fact cause the authority having jurisdiction to require an explosion-proof seal since non-explosion-proof seals may not be identified for this purpose.

"Identified" does not necessarily require the equipment to be listed by a qualified electrical testing laboratory, although the concept is included in the definition of the word in Article 100: "Identified (as applied to equipment). Recognizable as suitable for the specific purpose, function, use, environment, application, and so forth, where described in a particular Code requirement." The accompanying Fine Print Note indicates that an example of equipment being identified includes listing by a qualified electrical testing laboratory: "FPN: Some examples of ways to determine suitability of equipment for a specific purpose, environment, or application include investigations by a qualified testing laboratory (listing and labeling), an inspection agency, or other organizations concerned with product evaluation."

Seals in conduits passing from Division 2 locations into unclassified locations are needed to prevent the passage of gases or vapors, not to contain explosions in the conduit system, as is the case with Division 1 conduit systems. The standard for sealing fittings (UL 886) includes a pressure test of .007 cubic foot of air per hour at a pressure of 6 inches of water to ensure that passage of gas is minimized. All seals that have been evaluated to that standard are also explosion-proof. If other seals are found acceptable, but no specific requirements are provided, the authority having jurisdiction would have no basis for approval.

Electrical inspectors who are uncomfortable in making a determination about the capability of a conduit seal to "minimize passage of gases under normal operation" may in fact require that the conduit at the transition between a Class I, Division 2 and an unclassified area be a listed sealing fitting.

501.15(F) (3)

Sealing and Drainage, Canned Pumps, Process, or Service Connections, etc.

NEC page 353

CHANGE TYPE: Revision (Proposal 14-40; Comment 14-47)

CHANGE SUMMARY: NEC 501.15(F)(3) will now include two options for process sealing. Equipment listed and marked "Dual Seal" no longer requires additional sealing when used within manufacturer's ratings.

2005 CODE: "**510.15(F)(3) Canned Pumps, Process, or Service Connections, etc.** For canned pumps, process, or service connections for flow, pressure, or analysis measurement, and so forth, that depend on a single compression seal, diaphragm, or tube to prevent flammable or combustible fluids from entering the electrical raceway or cable system capable of transmitting fluids, an additional approved seal, barrier, or other means shall be provided to prevent the flammable or combustible fluid from entering the raceway or cable system capable of transmitting fluids beyond the additional devices or means, if the primary seal fails. The additional approved seal or barrier and the interconnecting enclosure shall meet the temperature and pressure conditions to which they will be subjected upon failure of the primary seal, unless other approved means are provided to accomplish the purpose. Drains, vents, or other devices shall be provided so that primary seal leakage will be obvious.

FPN: See also the fine print notes to 501.5.

Process-connected equipment that is listed and marked "Dual Seal" shall not require additional process sealing when used within the manufacturer's ratings.

FIGURE 5.5

FPN: For construction and testing requirements for dual-seal process-connected equipment, refer to ISA 12.27.01, *Requirements for Process Sealing Between Electrical Systems and Potentially Flammable or Combustible Process Fluids.*"

CHANGE SIGNIFICANCE: Section 501.15(F)(3) previously attempted to address the issue of process sealing of electrical equipment, but placed the burden of ensuring reliable sealing primarily on the installer. It is widely recognized in the industry that there are very few reliable means of affecting a process seal in the field. Some installers assume that a standard poured conduit seal is sufficient, yet it is clear that these seals are typically not rated or suitable for process conditions, which typically involve aggressive materials at high pressures and temperatures.

The ISA SP12 Committee and ISA SP 12.27 Subcommittee have been working for years to address industry concerns related to the reliability of process seals that form part of electrical equipment. The result of this effort is ISA 12.27.01, *Requirements for Process Sealing Between Electrical Systems and Potentially Flammable or Combustible Process Fluids.*

Following the lead of ISA 12.27.01, the intent of this proposal was to off-load the responsibility for effecting an adequate process seal from the installer and place it on the manufacturer of the process-connected equipment as long as the equipment has been examined, identified, and marked in accordance with ISA 12.27.01. The Code Panel stated that a single seal (which is what was proposed) is not considered by CMP-14 to afford an acceptable level of safety. However, the Panel also recognized that evaluation and verification of unlisted process seals is not easily accomplished in the field. Therefore, "identified" was changed to "listed" for dual-seal equipment.

The text will now allow two methods of connecting this kind of equipment. The previously provided method was if the canned pump equipment is provided with a single seal, another seal is required to be installed. The additional seal is required to "meet the temperature and pressure conditions to which they will be subjected upon failure of the primary seal." (This information should be obtained from the manufacturer.) An enclosure that is vented or is provided with a drain can be installed between the seal and the source of the branch circuit to indicate that failure of the seals.

501.25, 502.25, 503.25, 505.19

Uninsulated Exposed Parts, Class I, Divisions 1 & 2; Class II, Divisions 1 & 2; Class III, Divisions 1 & 2

NEC pages 353, 359, 364, 381

CHANGE TYPE: Revision (Proposal 14-47a; Comment 14-51)

CHANGE SUMMARY: CMP-14 has revised sections in Articles 501, 502, 503, and 505 due to changes in the definition of *live part* in the 2002 NEC.

2005 CODE: "**501.25 <u>Uninsulated Exposed</u> ~~Live~~ Parts, Class I, Divisions 1 & 2.** There shall be no <u>uninsulated</u> exposed ~~live~~ parts, <u>such as electric conductors, buses, terminals, or components, that operate at more than 30 volts (15 volts in wet locations). These parts shall additionally be protected by a protection technique according to 500.7(E), 500.7(F), or 500.7(G) that is suitable for the location.</u>"

"**502.25 <u>Uninsulated Exposed</u> ~~Live~~ Parts, Class II, Divisions 1 & 2.** <u>There shall be no uninsulated</u> ~~live parts shall not be~~ exposed <u>parts, such as electric conductors, buses, terminals, or components, that operate at more than 30 volts (15 volts in wet locations). These parts shall additionally be protected by a protection technique according to 500.7(E), 500.7(F), or 500.7(G) that is suitable for the location.</u>"

FIGURE 5.6

"**503.25 Uninsulated Exposed** ~~Live~~ **Parts, Class III, Divisions 1 & 2.** There shall be no uninsulated ~~Live parts shall not be~~ exposed parts, such as electric conductors, buses, terminals, or components, that operate at more than 30 volts (15 volts in wet locations). These parts shall additionally be protected by a protection technique according to 500.7(E), 500.7(F), or 500.7(G) that is suitable for the location.

Exception: As provided in 503.155."

"**505.19 Uninsulated Exposed** ~~Live~~ **Parts.** There shall be no uninsulated exposed parts, such as electric conductors, buses, terminals, or components that operate at more than 30 volts (15 volts in wet locations). These parts shall additionally be protected by type of protection ia, ib, or nA that is suitable for the location."

CHANGE SIGNIFICANCE: The Code Panel substantiation for this proposal stated "The 2002 NEC changed the definition of *live part* from 'electric conductors, buses, terminals, or components that are uninsulated or exposed and a shock hazard exists' to 'energized conductive components.' This definition change serves to prohibit the use of exposed low voltage/intrinsically safe circuits that are not a shock hazard or an ignition hazard. This change uses the wording of the original definition and also calls attention to the need for protection from ignition of combustible materials in addition to protection from electric shock."

The Panel action has quantified the voltage level that presents a shock hazard for Class I, Class II, and Class III locations. The Panel selected the word *exposed* to qualify the equipment to which the revised rules apply. "Exposed (as applied to live parts)" is defined in Article 100 as, "Capable of being inadvertently touched or approached nearer than a safe distance by a person. It is applied to parts that are not suitably guarded, isolated, or insulated."

The section reference for 501.25 was 501.15 in the 2002 NEC; 502.25 was 502.15 and 503.25 was 503.15.

502.15(4)

Sealing, Class II, Divisions 1 and 2

NEC page 359

CHANGE TYPE: Revision (Proposal 14-57; Comment 14-60)

CHANGE SUMMARY: A new item (4) has been added to the list of permitted methods of providing a seal or an equivalent to a sealing fitting for Class II, Divisions 1 and 2 locations.

2005 CODE: "502.15 Sealing, Class II, Divisions 1 and 2. Where a raceway provides communication between an enclosure that is required to be dust-ignition proof and one that is not, suitable means shall be provided to prevent the entrance of dust into the dust-ignition-proof enclosure through the raceway. One of the following means shall be permitted:

(1) A permanent and effective seal

(2) A horizontal raceway not less than 3.05 m (10 ft) long

(3) A vertical raceway not less than 1.5 m (5 ft) long and extending downward from the dust-ignitionproof enclosure

(4) <u>A raceway installed in a manner equivalent to (2) or (3) that extends only horizontally and downward from the dust-ignitionproof enclosure</u>

Where a raceway provides communication between an enclosure that is required to be dust-ignitionproof and an enclosure in an unclassified location, seals shall not be required.

Sealing fittings shall be accessible.

Seals shall not be required to be explosionproof.

FPN: "Electrical sealing putty is a method of sealing."

CHANGE SIGNIFICANCE: Many installations require conduits to be installed other than directly down or horizontally from the dust-ignition-proof enclosure because of obstructions of various kinds (including other parts of the electrical installation). For example, a conduit may run 5 feet horizontally and then turn downward for 5 feet or more. Such an installation provides for equal or better prevention of dust migration into an enclosure, yet it would not be recognized under the 2002 NEC as effective because it is not 10 feet in the horizontal run and extends horizontally rather than downward from the enclosure itself.

Consider two more examples that illustrate the weaknesses of the previous rule. If a conduit run extends 1 foot horizontally and then 10 feet down, it must have a seal, but a conduit extending 5 feet down and then 1 foot horizontally is considered equivalent to a seal. Similarly, if a conduit extends 1 foot down and then 20 feet horizontally, it must have a seal, but another conduit extending 10 feet horizontally and then 1 foot down is considered equivalent to a seal. A reasonable interpretation may allow such installations, but many interpretations rely only on the literal wording of the Code and not on the performance of the installation.

The current language, accepted by CMP-14, should clear up the application of providing seals or seal equivalent for these installations.

The length of the conduit in the horizontal position allows dust particles suspended in the air to settle, as they are heavier than air. But in the vertical position the heavier-than-air particles would not rise far enough to enter the enclosure.

300 mm (12 in.)

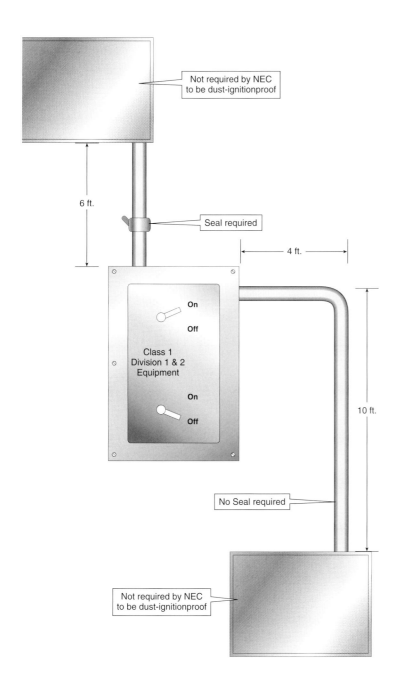

Not required by NEC to be dust-ignitionproof

6 ft.

Seal required

4 ft.

On
Off

Class 1
Division 1 & 2
Equipment

On
Off

10 ft.

No Seal required

Not required by NEC to be dust-ignitionproof

FIGURE 5.7

502.10(A)(2)

Class II, Division 1, Flexible Connections

NEC page 359

CHANGE TYPE: Revision (Proposal 14-55; Comment 14-62)

CHANGE SUMMARY: Jacketed Type MC cable is now permitted to be installed for flexible connections in Class II, Division 1 locations.

2005 CODE: "**502.10(A) Class II, Division 1.**

(1) **General.** In Class II, Division 1 locations, the wiring methods in (1) through (4) shall be permitted.

(1) Threaded rigid metal conduit, or threaded steel intermediate metal conduit.

(2) Type MI cable with termination fittings listed for the location. Type MI cable shall be installed and supported in a manner to avoid tensile stress at the termination fittings.

(3) In industrial establishments with limited public access, where the conditions of maintenance and supervision ensure that only qualified persons service the installation, Type MC cable, listed for use in Class II, Division 1 locations, with a gas/vaportight continuous corrugated metallic sheath, an overall jacket of suitable polymeric material, separate grounding conductors in accordance with 250.122, and provided with termination fittings listed for the application, shall be permitted.

(4) Fittings and boxes shall be provided with threaded bosses for connection to conduit or cable terminations and shall be dust-tight. Fittings and boxes in which taps, joints, or terminal connections are made, or that are used in Group E locations, shall be identified for Class II locations.

FIGURE 5.8

(2) Flexible Connections. ~~(c)~~ Where necessary to employ flexible connections, <u>one or more of the following shall be permitted:</u>

<u>(1)</u> Dust-tight flexible connectors

<u>(2)</u> Liquid-tight flexible metal conduit with listed fittings

<u>(3)</u> Liquid-tight flexible non-metallic conduit with listed fittings

<u>(4) Interlocked armor Type MC cable having an overall jacket of suitable polymeric material and provided with termination fittings listed for Class II, Division 1 locations</u>

(5) Flexible cord listed for extra-hard usage and provided with bushed fittings ~~shall be used~~. Where flexible cords are used, they shall comply with 502.140. ~~Where flexible connections are subject to oil or other corrosive conditions, the insulation of the conductors shall be of a type listed for the condition or shall be protected by means of a suitable sheath.~~

FPN: "See 502.30(B) for grounding requirements where flexible conduit is used."

CHANGE SIGNIFICANCE: This revision indicates that interlocked armor Type MC cables are flexible by construction and, with an impervious outer jacket, are suitable for the applications provided in this section. Since three varieties of Type MC cable are produced and the smooth sheath and corrugated types may not be suitable where flexibility is needed, the interlocked armor Type MC cable is designated.

The proposal also modified the list format, which should be more user friendly.

503.10(A) (3)

Class III Wiring Methods, Nonincendive Field Wiring

NEC page 364

CHANGE TYPE: New (Proposal 14-63)

CHANGE SUMMARY: The wiring methods in Class III, Division 1 locations are very similar to those found in Class II, Division 2 locations. Nonincendive field wiring is now acceptable in Class III, Division 1 locations.

2005 CODE: "**(3) Nonincendive Field Wiring.** Nonincendive field wiring shall be permitted using any of the wiring methods permitted for unclassified locations. Nonincendive field wiring systems shall be installed in accordance with the control drawings(s). A simple apparatus, not shown on the control drawing, shall be permitted in a non-incendive field wiring circuit provided the simple apparatus does not interconnect the nonincendive field wiring circuit to any other circuit.

FPN: Simple apparatus is defined in 504.2.

Separate nonincendive field wiring circuits shall be installed in accordance with one of the following:

(1) In separate cables
(2) In multiconductor cables where the conductors of each circuit are within a grounded metal shield
(3) In multiconductor cables, where the conductors of each circuit have insulation with a minimum thickness of 0.25 mm (0.01 in.)"

CHANGE SIGNIFICANCE: The requirements for nonincendive field wiring in Class I and Class II locations were modified and expanded in the 2002 Code cycle, but no similar requirements were included for Class III locations. This proposal adds provisions for nonincendive wiring for Class III, Division 1 areas. Section 503.10(B) indicates that the wiring methods in (A), as well as some additional methods provided in the exception, are acceptable for use in Division 2 areas.

FIGURE 5.9

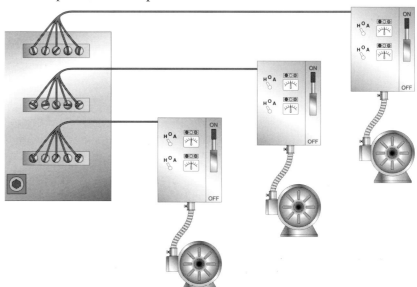

CHANGE TYPE: New (Proposal 14-108a)

CHANGE SUMMARY: This Proposal adds a new Article 506 to provide the Zone concept as an alternate method of addressing concerns for electrical installations in atmospheres where fire and explosion hazards may exist due to flammable dusts, fibers, or flyings. This revision provides an alternative to the requirements located in Articles 500, 502, and 503 for electrical installations in these environments.

2005 CODE: See the NEC for the Article.

ARTICLE **506 Zone 20, 21, and 22 Locations for Combustible Dusts, Fibers, and Flyings**

"**506.1 Scope.** This article covers the requirements for the Zone classification system as an alternative to the division classification system covered in Article 500, Article 502, and Article 503 for electrical and electronic equipment and wiring for all voltages in Zone 20, Zone 21, and Zone 22 hazardous (classified) locations where fire and explosion hazards may exist due to combustible dusts, or ignitable fibers, or flyings. Combustible metallic dusts are not covered by the requirements of this article.

Article 506

Zone 20, 21, and 22 Locations for Combustible Dusts, Fibers, and Flyings
NEC page 383

FIGURE 5.10

FPN NO. 1:   For the requirements for electrical and electronic equipment and wiring for all voltages in Class I, Division 1 or Division 2; Class II, Division 1 or Division 2; Class III, Division 1 or Division 2; and Class I, Zone 0, Zone 1, or Zone 2 hazardous (classified) locations where fire or explosion hazards may exist due to flammable gases or vapors, flammable liquids, or combustible dusts or fibers, refer to Articles 500 through 505.

FPN 2:   Zone 20, Zone 21, and Zone 22 area classifications are based on the modified IEC area classification system as defined in ISA 12.10.05, *Electrical Apparatus for Use In Zone 20, Zone 21, and Zone 22 Hazardous (Classified) Locations—Classification of Zone 20, Zone 21, and Zone 22 Hazardous (Classified) Locations (IEC61241-10 Mod).*

FPN 3:   "The unique hazards associated with explosives, pyrotechnics, and blasting agents are not addressed in this article."

CHANGE SIGNIFICANCE: A Zone based classification system for Zone 20, Zone 21, and Zone 22 for flammable dusts, fibers, and flyings was introduced in Europe earlier. Area classification standards and methods of protection have been defined for these areas, and standards have been developed and published by IEC TC 31, which is the International Electrotechnical Commission (IEC) committee that develops standards for electrical equipment in hazardous areas.

The United States participated in the development of these standards, and has recently joined the IEC Ex organization, whose goal is to eventually harmonize hazardous area equipment standards worldwide.

As was the case with the Zone classification system for flammable gases, installation rules and regulations also need to be in place in the United States for companies that wish to utilize equipment employing the Zone-based methods of protection. The Zone dust classification and protection methods therefore needed to be recognized by the NEC in order for installation of such equipment to be legal in many jurisdictions. The IEC zone system employed in Europe has been proven to be a safe system.

This revision allows the flexibility to use an alternate method that is recognized as safe. It provides the necessary framework to allow the recognition of the Zone dust protection system in the NEC by the addition of Article 506 to cover installation of electrical equipment in Zones 20, 21, and 22.

CHANGE TYPE: Revision (Proposal 14-111; Comments 14-98 and 14-99)

CHANGE SUMMARY: NEC 511.3 has been reorganized so that all text related to unclassified locations within commercial garages is now located in 511.3(A) and all text related to classified locations is now located in 511.3(B). Some of the text in past editions was included as exceptions, some was located in Article 514, and some was already in 511.3(A) or (B).

2005 CODE: See the following table for the organization of 511.3. Also, see the footnotes to the table.

511.3

Commercial Garages, Repair and Storage, Classification of Locations
NEC page 388

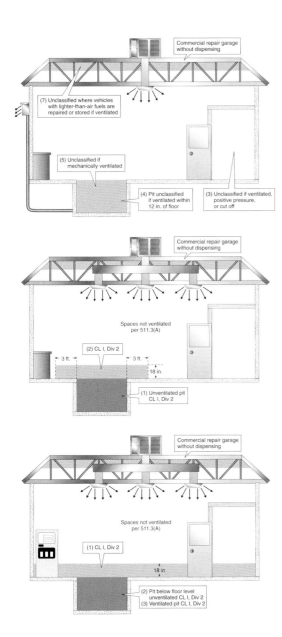

FIGURE 5.11

511.3(A) Unclassified Locations	511.3(B) Classified Locations
(1) Parking and repair garages[1]	(1) Flammable fuel dispensing areas
(2) Alcohol-based windshield washer fluid	(2) Lubrication or service room where Class I liquids or gaseous fuels (such as natural gas, hydrogen, or LPG) are **not** transferred
(3) Specific areas adjacent to classified locations[2]	(3) Lubrication or service room where Class I liquids or gaseous fuels (such as natural gas, hydrogen, or LPG) **are** transferred
(4) Pits in lubrication or service room where Class liquids are not transferred[3]	(4) Within 450 mm (18 in.) of the ceiling when repairing vehicles with lighter-than-air fuels
(5) Up to a level of 450 mm (18 in.) above the floor in lubrication or service rooms where Class I liquids are transferred[4]	
(6) Flammable liquids having flash points below 38°C (100°F) not transferred[5]	
(7) Within 450 mm (18 in.) of the ceiling not classified for lighter-than-air fuels[6]	

[1] Are unclassified if designed per 511.3(A)(2) through (A)(7)

[2] Are classified without ventilation, positive pressure, or cut-off

[3] Are classified without exhaust ventilation within 12 in. of floor

[4] Is classified unless ventilated with 4 air changes per hour

[5] Is classified if fuels are transferred

[6] Is classified unless ventilated at 1 cfm/sq ft

CHANGE SIGNIFICANCE: This proposal reorganized the existing requirements found in 511.3 and Table 514.3(B)(1). The reorganization places all unclassified spaces in commercial repair and storage garages into Subsection (A) of 511.3 and all classified spaces into Subsection (B). It should be noted that the "unclassification" of the locations or situations referred to in (3), (4), (5), (6), and (7) in the unclassified column is conditional on exhaust, ventilation, positive air pressure, or being cut off. All of these spaces would be under the classified column if the exhaust or ventilation did not occur at the specified rate, or if the space was not positively pressurized or is not cut off.

Some of the unclassified spaces were previously found in Subsection (A), while others were located as exceptions to the classified spaces; some classifications are extracted from Article 514. Some of the classified spaces of commercial garages were located in the previous Subsection (B), while others are from Article 514.

The proposal intended to correlate the existing extracted electrical requirements (in accordance with the NEC regulations for extraction of requirements from other NFPA standards/codes) for repair garages from the previous, outdated editions of NFPA 88A, 88B, and 30A with the most recent and revised editions of NFPA 88A and NFPA 30A. Please note that NFPA 88B requirements have been incorporated into the latest edition of NFPA 30A, NFPA 88B was withdrawn as a standard at the 2002 NFPA Fall Meeting in Atlanta.

CHANGE TYPE: Deletion (Proposal 14-115)

CHANGE SUMMARY: The previous 511.4(A)(1) and exception have been deleted. This eliminates the text that indicated that the space within a slab or masonry wall or below the slab was a Class I location in some conditions.

2005 CODE: "**511.4 Wiring and Equipment in Class I Locations.**
(A) Wiring Located in Class I Locations.** Within Class I locations as classified in 511.3, wiring shall conform to applicable provisions of Article 501.

~~(1) Raceways. Raceways embedded in a masonry wall or buried beneath a floor shall be considered to be within the Class I location above the floor if any connections or extensions lead into or through such areas.~~

~~Exception: Rigid nonmetallic conduit that complies with Article 352 shall be permitted where buried under not less than 600 mm (24 in.) of cover. Where rigid nonmetallic conduit is used, threaded rigid metal conduit or threaded steel intermediate metal conduit shall be used for the last 600 mm (24 in.) of the underground run to emergence or to the point of connection to the aboveground raceway and an equipment grounding conductor shall be included to provide electrical continuity of the raceway system and for grounding of non-current-carrying metal parts.~~"

CHANGE SIGNIFICANCE: The definition of a Class I location in Article 500 requires that an ignitable concentration of fuel and air be present for a classified location to exist. It was determined that, unless there was a void in these areas where air could collect, enough oxygen to create an ignitable concentration would not be available underground.

Now the applicable rules for wiring and sealing, which are located in Article 501, will apply to raceways entering classified locations from unclassified areas within or below a slab.

511.4

Commercial Repair Garage, Wiring and Equipment in Class I Locations
NEC page 390

FIGURE 5.12

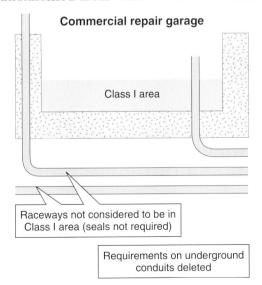

Commercial repair garage

Class I area

Raceways not considered to be in Class I area (seals not required)

Requirements on underground conduits deleted

513.12

Aircraft Hangars, Ground-Fault Circuit-Interrupter Protection for Personnel

NEC page 393

CHANGE TYPE: New (Proposal 14-132)

CHANGE SUMMARY: This new section adds ground-fault circuit-interrupter (GFCI) protection for 15- and 20-ampere, 125-volt, single-phase, 50/60-Hz receptacles that are installed in aircraft hangars to serve diagnostic equipment, electrical hand tools, or portable lighting equipment.

2005 CODE: "**513.12 Ground-Fault Circuit-Interrupter Protection for Personnel.** All 125-volt, 50/60-Hz, single-phase, 15- and 20-ampere receptacles installed in areas where electrical diagnostic equipment, electrical hand tools, or portable lighting equipment are to be used shall have ground-fault circuit-interrupter protection for personnel."

CHANGE SIGNIFICANCE: Personnel servicing and maintaining aircraft use the same hand tools and equipment that are used in commercial garages and should be afforded the same level of personnel protection. It is also common to have concrete floors that are on grade and considered a grounded surface. Very often, service or repair personnel use extension cords and portable tools in their work—the risk of electric shock is quite high.

To avoid confusion, CMP-14 limited the requirement for GFCI protection to 50/60-Hz devices. Many hangars include 400-Hz equipment, which were determined not to be compatible with GFCI protection.

FIGURE 5.13

CHANGE TYPE: Revision (Proposals 14-134 and 14-136)

CHANGE SUMMARY: The definition of *motor fuel dispensing facility* has been revised to include marine craft. Also, a new footnote extracted from NFPA 30A has added to Table 514.3(B)(1) to provide a definition of grade level where gasoline dispensers are installed on over-water piers at marinas.

2005 CODE: "**514.2 Motor Fuel Dispensing Facility.** That portion of a property where motor fuels are stored and dispensed from fixed equipment into A location where gasoline or other volatile flammable liquids or liquefied flammable gases are transferred to the fuel tanks (including auxiliary fuel tanks) of motor self-propelled vehicles or marine craft or into approved containers, including all equipment used in connection therewith. [NFPA 30A:3.3.11]

FPN: Refer to Articles 510 and 511 with respect to electric wiring and equipment for other areas used as lubritoriums, service rooms, repair rooms, offices, salesrooms, compressor rooms, and similar locations."

Table 514.3(B)(1) Class I Locations —Motor Fuel
Dispensing Facilities

Location	Class I, Group, Division	Extent of Classified Location[1]
(Table Items)	(Table Items)	(Table Items)
(Table Items)	(Table Items)	(Table Items)

[1] For marine applications, *grade level* means the surface of a pier extending down to water level.

CHANGE SIGNIFICANCE: The definition of *motor fuel dispensing facility* has been changed to agree with the definition in NFPA 30A. The information in the brackets below the definition identify it as extracted from a parent NFPA Code or standard. NFPA 30A is the *Code for Motor Fuel Dispensing Facilities and Repair Garages.* As can be seen, the fueling of "marine craft" has been added to the definition and

514.2, Table 514.3(B) (1)

Motor Fuel Dispensing Facility, Table 514.3(B)(1) Class I Locations—Motor Fuel Dispensing Facilities
NEC pages 394, 395

FIGURE 5.14

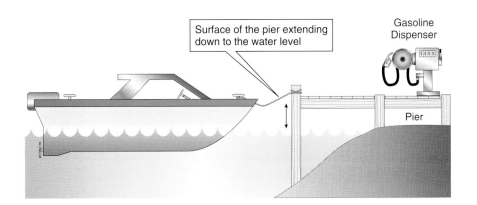

Surface of the pier extending down to the water level

Gasoline Dispenser

Pier

so that operation is required to comply with Article 514. Additional requirements for fueling of marine craft can be found in 555.20.

The footnote added to Table 514.3(B)(1) provides enforceable text in the NEC for local enforcement authorities who only adopt the NEC and do not adopt other referenced standards, such as NFPA 30A.

Prior to the 1975 edition, Article 555 was silent on (gasoline) dispensing stations at a marina, specifically dispensing equipment located on an over-water pier. Proposal 11 in the Preprint of Proposed Amendments for the 1974 NEC added requirements at 555-9. Submitted by the NEC Technical Subcommittee on Marinas and Boatyards, the text spelled out where the lower limit of the hazardous location was for marine applications; this was set at the lowest water surface. This was a necessary change, since for a marine application, there isn't a "grade level" referenced to earth as is the case for a dispenser located on shore.

In the 1981 NEC, the text delineating Class I locations in 514-2 was deleted and Table 514-2 was created using Table 7-1 from NFPA 30. The new table did not contain any reference for marine applications. In 1984, the chapter covering automotive and marine service stations was deleted from NFPA 30 and used as the basis for a new document, NFPA 30A.

For the 1990 NEC, 555-9 was revised to delete the area classifications and merely require compliance with Article 514. The Fine Print Note was modified to include references to NFPA 30A and NFPA 303. This appears to have been done to comply with a NEC Correlating Committee note regarding extracted material. The net result of this was to leave the NEC without any reference to the lowest water surface for marine applications, a vital piece of information for proper application of the area classification. Part of the substantiation in 1981 to include Table 514-2 was to provide information without the need for an additional reference code. However, with the deletion in 555-9, Code users needed to go to NFPA 30A to find the requirements. (Many states do not adopt the referenced standards in Fine Print Notes since these notes are informational only and not enforceable as requirements.) Adding this new footnote, which is taken from Table 8.3.1 of NFPA 30A, will place the extent of these classified locations into the NEC as an enforceable requirement in the same manner as a shore-based dispenser.

While this addition to the table clarifies the grade level for gasoline dispensers located on piers, there is little help for the designer, installer, or inspector to apply the rules of Table 514.3(B)(1) to gasoline dispensers located on floating docks. Yet where the dock is of open construction, such as beams and a deck supported by pontoons, it is probably prudent to apply the area classification from Table 514.3(B)(1).

Independent judgment will need to be exercised on determining the extent of hazardous locations where the floating dock is of closed construction, such as lightweight concrete over a foam core. Such docks will usually have a concrete deck with PVC pipe sleeves and deck boxes cast into the construction. Gasoline spills can easily enter these deck boxes, and vapor or liquid can travel for considerable distances. Wiring and other services such as water, telephone, cable and fuel supply piping are run through these chases below the deck.

CHANGE TYPE: Revision (Proposal 14-122)

CHANGE SUMMARY: The space below a Class I, Division 1 or 2 location in the vicinity of a gasoline dispenser is no longer considered a Class I, Division 1 location.

2005 CODE: "**514.8 Underground Wiring.** Underground wiring shall be installed in threaded rigid metal conduit or threaded steel intermediate metal conduit. Any portion of electrical wiring ~~or equipment~~ that is below the surface of a Class I, Division 1, or a Class I, Division 2 location [as classified in Table 514.3(B)(1) and Table 514.3(B)(2)] shall be <u>sealed within 3.05 m (10 ft) of</u> ~~considered to be in a Class I, Division 1, location that shall extend at least to~~ the point of emergence above grade. <u>Except for listed explosion-proof reducers at the conduit seal, there shall be no union, coupling, box, or fitting between the conduit seal and the point of emergence above grade.</u> Refer to Table 300.5.

Exception No. 1: Type MI cable shall be permitted where it is installed in accordance with Article 332.

Exception No. 2: Rigid nonmetallic conduit ~~complying with Article 352~~ *shall be permitted where buried under not less than 600 mm (2 ft) of cover. Where rigid nonmetallic conduit is used, threaded rigid metal conduit or threaded steel intermediate metal conduit shall be used for the last 600 mm (2 ft) of the underground run to emergence or to the point of connection to the aboveground raceway, and an equipment grounding conductor shall be included to provide electrical continuity of the raceway system and for grounding of non-current-carrying metal parts.*"

514.8

Motor Fuel Dispensing Facilities, Underground Wiring
NEC page 394

FIGURE 5.15

CHANGE SIGNIFICANCE: This proposal was submitted by the Technical Committee on Automotive and Marine Service Stations, 30A. Much of the material in Article 514 is extracted from NFPA 30A.

The deletion to the reference to equipment takes into account that this section is for underground wiring, not equipment. Electrical classification of equipment is already covered under Table 514(3)(B)(1), which is the extracted NFPA 30A Table 8.3.1.

The NFPA 30A Committee notes that the space below the surface of a Class I, Division 1 or 2 location does not meet the literal definition in NEC Section 500.5(B)(1) of a Class I, Division 1 location, unless a pocket or void is left in the earth to collect air. However, the revised text still requires any electrical wiring that is installed in the ground under Class I, Division 1 or 2 locations to be sealed in accordance with 501.15(A). Therefore, the end result is the same requirement for underground electrical wiring under Section 514.8, but without the technically incorrect electrical classification of the earth as a Class I, Division 1 or 2 location. Also, conduits now must be sealed within 10 feet of point of emergence above grade.

516.3
Classification of Locations
NEC page 402

CHANGE TYPE: Revision (Proposal 14-145a; Comment 14-115)

CHANGE SUMMARY: These revisions made to Article 516 include the Zone concept for spray areas.

2005 CODE: "**516.3 Classification of Locations.** Classification is based on dangerous quantities of flammable vapors, combustible mists, residues, dusts, or deposits.

(A) **Class I, Division 1 or Class I, Zone 0 Locations.** The following spaces shall be considered Class I, Division 1, or Class I, Zone 0, as applicable:

(1) The interior of any open or closed container of a flammable liquid

(2) The interior of any dip tank or coating tank

FPN: For additional guidance and explanatory diagrams, see 4.3.5 of NFPA 33-2003, *Standard for Spray Application Using Flammable or Combustible Materials,* and Sections 4.2, 4.3, and 4.4 of NFPA 34-2003, *Standard for Dipping and Coating Processes Using Flammable or Combustible Liquids.*

(B) **Class I or Class II, Division 1 Locations.** The following spaces shall be considered Class I, Division 1; Class I, Zone 1; or Class II, Division 1 locations, as applicable:

(1) The interior of spray booths and rooms except as specifically provided in 516.3(D)

(2) The interior of exhaust ducts

(3) Any area in the direct path of spray operations

FIGURE 5.16

(4) For <u>open</u> dipping and coating operations, all space within a 1.5-m (5-ft) radial distance from the vapor sources extending from these surfaces to the floor. The vapor source shall be the liquid exposed in the process and the drainboard, and any dipped or coated object from which it is possible to measure vapor concentrations exceeding 25 percent of the lower flammable limit at a distance of 300 mm (1 ft), in any direction, from the object.

(5) Sumps, pits, or below-grade channels within 7.5 m (25 ft) horizontally of a vapor source. If the sump, pit, or channel extends beyond 7.5 m (25 ft) from the vapor source, it shall be provided with a vapor stop or it shall be classified as Class I, Division 1 for its entire length.

(6) <u>All space in all directions outside of but within 900 mm (3 ft) of open containers, supply containers, spray gun cleaners, and solvent distillation units containing flammable liquids.</u> ~~The interior of any enclosed dipping or coating process or apparatus. [NFPA 33, 1.6 Definitions; NFPA 34, 1.6, Definitions, 4.2.1, 4.2.2, 4.3.1]~~

<u>(C)</u> ~~(B)~~ Class I or Class II, Division 2 Locations. The following spaces shall be considered Class I<u>, Division 2 or Class I, Zone 2;</u> or Class II, Division 2 as applicable.

(1) Open Spraying. For open spraying, all space outside of but within 6 m (20 ft) horizontally and 3 m (10 ft) vertically of the Class I, Division 1 <u>or Class I, Zone 1</u> location as defined in 516.3(A), and not separated from it by partitions. See Figure 516.3(B)(1) [NFPA 33:6.5.1].

(2) Closed-Top, Open-Face, and Open-Front Spraying. If spray application operations are conducted within a closed-top, open-face, or open-front booth or room, any electrical wiring or utilization equipment located outside of the booth or room but within the boundaries designated as Division 2 <u>or Zone 2</u> in Figure 516.3(B)(2) shall be suitable for Class I, Division 2<u>; Class I, Zone 2;</u> or Class II, Division 2 locations, whichever is applicable. The Class I, Division 2<u>; Class I, Zone 2;</u> or Class II, Division 2 locations shown in Figure 516.3(B)(2) shall extend from the edges of the open face or open front of the booth or room in accordance with the following:

(a) If the exhaust ventilation system is interlocked with the spray application equipment, ~~then~~ the Division 2 location shall extend 1.5 m (5 ft) horizontally and 900 mm (3 ft) vertically from the open face or open front of the booth or room, as shown in Figure 516.3(B)(2), top.

(b) If the exhaust ventilation system is not interlocked with the spray application equipment, then the Division 2 <u>or Zone 2</u> location shall extend 3 m (10 ft) horizontally and 900 mm (3 ft) vertically from the open face or open front of the booth or room, as shown in Figure 516.3(B)(2), bottom.

For the purposes of this subsection, *interlocked* shall mean that the spray application equipment cannot be operated unless the exhaust ventilation system is operating and functioning properly and spray application is automatically stopped if the exhaust ventilation system fails. [NFPA 33:6.5.2]

(3) Open-Top Spraying. For spraying operations conducted within an open-top spray booth, the space 900 mm (3 ft) vertically above the booth and within 900 mm (3 ft) of other booth openings shall be considered Class I, <u>Division 2; Class I, Zone 2;</u> or Class II, Division 2. [NFPA 33:6.5.3]

(4) Enclosed Booths and Rooms. For spraying operations confined to an enclosed spray booth or room, the space within 900 mm (3 ft) in all directions from any openings shall be considered Class I, <u>Division 2; Class I or Zone 2;</u> or Class II, Division 2 as shown in Figure 516.3(B)(4). [NFPA 33:6.5.4]

(5) Dip Tanks and Drain Boards—Surrounding Space. For dip tanks and drain boards, the 914-mm (3-ft) space surrounding the Class I, Division 1 <u>or Class I, Zone 1</u> location as defined in 516.3(A)(4) and as shown in Figure 516.3(B)(5). [NFPA 34:6.4.3]

(6) Dip Tanks and Drain Boards—Space Above Floor. For dip tanks and drain boards, the space 900 mm (3 ft) above the floor and extending 6 m (20 ft) horizontally in all directions from the Class I, Division 1, <u>or Class I, Zone 1</u> location.

Exception: This space shall not be required to be considered a hazardous (classified) location where the vapor source area is 0.46 m² (5 ft²) or less, and where the contents of the open tank trough or container do not exceed 19 L (5 gal). In addition, the vapor concentration during operation and shutdown periods shall not exceed 25 percent of the lower flammable limit outside the Class I location specified in 516.3(A)(4). [~~NFPA 33; 4.3.1, 4.3.2, 4.3.3, 4.3.4~~; NFPA 34:<u>6.4.4</u> ~~, 4.2.3, 4.2.4~~]

<u>**(7) Open Containers.** All space in all directions within 600 mm (2 ft) of the Division 1 or Zone 1 area surrounding open containers, supply containers, spray gun cleaners, and solvent distillation units containing flammable liquids, as well as the area extending 1.5 m (5 ft) beyond the Division 1 or Zone 1 area up to a height of 460 mm (18 in.) above the floor or grade level. [NFPA 33:6.5.5.1(2)]</u>"

CHANGE SIGNIFICANCE: This proposal introduces the Zone concept as an option for protection of electrical systems in Article 516. The Zone concept of hazardous location wiring was introduced into the 1996 Code as Article 505. The concept continues to be expanded upon in each edition of the NEC, as can be seen by the new Article 506 for Zone concept wiring in traditional Class II and Class III areas. Before this update, the Zone concept of hazardous location wiring had previously been used for many years in European countries and other countries that adopted the International Electrotechnical Committee (IEC) codes and standards, including Canada.

517.13 (A), (B)

Grounding of Receptacles and Fixed Electric Equipment in Patient Care Areas

NEC page 411

CHANGE TYPE: Revision (Proposal 15-17)

CHANGE SUMMARY: Changes to this section on wiring methods required in patient care areas of health care facilities include some editorial work. Also, 250.118 should now be used to determine the wiring methods suitable as an equipment grounding conductor.

2005 CODE: "**517.13 Grounding of Receptacles and Fixed Electric Equipment in Patient Care Areas.** Wiring in patient care areas shall comply with 517.13(A) and (B).

(A) **Wiring Methods.** All branch circuits serving patient care areas shall be provided with a ground path for fault current by installation in a metal raceway system, or a cable <u>having a metallic </u>armor or sheath assembly. The metal raceway system, or <u>metallic </u>cable armor or sheath assembly, shall itself qualify as an equipment grounding return path in accordance with 250.118. ~~Type AC, Type MC, Type MI cables shall have an outer metal armor or sheath that is identified as an acceptable grounding return path.~~

(B) **Insulated Equipment Grounding Conductor.** ~~In an area used for patient care,~~ The grounding terminals of all receptacles and all non-current-carrying conductive surfaces of fixed electric equipment likely to become energized that are subject to personal contact, operating at over 100 volts, shall be grounded by an insulated copper conductor. The equipment grounding conductor shall be sized in accordance with Table 250.122 and installed in metal raceways or ~~metal-clad~~ as a part of listed cables having a metallic armor or sheath assembly with the branch-circuit conductors supplying these receptacles or fixed equipment.

FIGURE 5.17

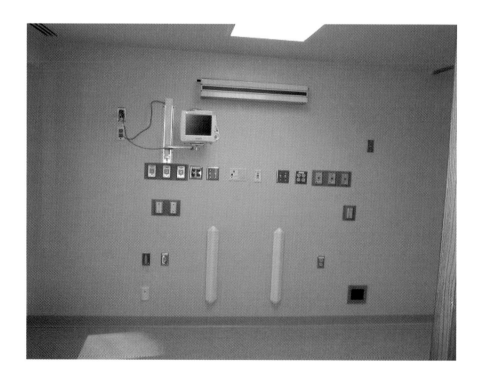

Exception No. 1: Metal faceplates shall be permitted to be grounded by means of a metal mounting screw(s) securing the faceplate to a grounded outlet box or grounded wiring device.

Exception No. 2: Luminaires (light fixtures) more than 2.3 m (7$^1/_2$ ft) above the floor and switches located outside of the patient vicinity shall not be required to be grounded by an insulated equipment grounding conductor."

CHANGE SIGNIFICANCE: This proposal was intended first as an editorial improvement. Several redundant words or phrases were eliminated. The revision also permits Section 250.118 to control which cable wiring systems are suitable as an equipment grounding conductor. Cable wiring systems will now be treated equal to metal raceways. Section 250.118(5) permits flexible metal conduit that is not listed for grounding to be used as an equipment grounding conductor for no more than 6 feet. Section 250.118(6) permits listed liquidtight flexible metal conduit (not listed for grounding) to be used as an equipment grounding conductor in lengths up to 6 feet in the ground return path, in some cases with contained conductors protected by no more than a 60-ampere OC device.

The following table summarizes the armored cable types that are or are not permitted to be used for wiring in patient care areas. Wiring methods requirements must be correlated with the requirements for physical protection of the emergency system in hospitals, found in 517.30(C)(3). No such requirements exist for other health care facilities, such as nursing homes, doctor and dentist offices, limited care facilities, or outpatient surgery centers, unless they have an essential electrical system that is required to comply with 517.30 through 517.35.

Cable Type	Acceptable for Wiring in Patient Care Area?
Type AC with an insulated equipment grounding conductor	Yes
Interlocked-armor Type MC with one or more equipment grounding conductors	No
Interlocked-armor Type MC with an internal grounding/bonding conductor and an insulated equipment grounding conductor	Yes
Corrugated-Tube Type MC cable with a bare equipment grounding conductor with or without an insulated equipment grounding conductor	No
Corrugated-Tube Type MC Cable without a bare equipment grounding conductor but with an insulated equipment grounding conductor	Yes

517.17(A)

Health Care Facilities, Ground-Fault Protection

NEC page 411

CHANGE TYPE: Revision (Proposal 15-24a; Comment 15-14)

CHANGE SUMMARY: The equipment ground-fault protection requirements apply "to hospitals and other buildings [housing] critical care areas or utilizing life support equipment, and buildings that provide the required essential utilities or services for the operation of critical care areas or electrical life support equipment."

2005 CODE: "(A) Applicability. The requirements of 517.17 shall apply to hospitals and other buildings (including multiple occupancy buildings) with critical care areas or utilizing electrical life support equipment, and buildings that provide the required essential utilities or services for the operation of critical care areas or electrical life support equipment.

(B) (A) Feeders. Where ground-fault protection is provided for operation of the service disconnecting means or feeder disconnecting means as specified by 230.95 or 215.10, an additional step of ground-fault protection shall be provided in all the next level of feeder disconnecting means downstream toward the load. Such protection shall consist of overcurrent devices and current transformers or other equivalent protective equipment that shall cause the feeder disconnecting means to open.

The additional levels of ground-fault protection shall not be installed as follows:

(1) On the load side of an essential electrical system transfer switch

(2) Between the on-site generating unit(s) described in 517.35(B) and the essential electrical system transfer switch(es)

FIGURE 5.18

(3) On electrical systems that are not solidly grounded wye systems with greater than 150 volts to ground but not exceeding 600 volts phase-to-phase"

CHANGE SIGNIFICANCE: Several proposals were submitted for this section to try to clarify when a second level of ground-fault protection should be appropriately required. Panel 15 (formerly Panel 17 for many NEC cycles) determined that language on the applicability of the requirement was needed to clearly identify when and where a second level of ground-fault protection is needed to prevent nuisance tripping of the main. Panel 15 suggested that a Fine Print Note be added to 230.95 and 215.10, referencing the special requirements of 517.17 for all structures that may potentially contain health care occupancies.

The previous wording in this section reportedly caused many problems for enforcement officials and theoretically makes no sense. The section previously required that all health care facilities have a second level of ground-fault protection where service or feeder ground-fault protection is provided. For example, a strip mall with service equipment ground-fault protection and minor medical or dental occupancies in addition to retail occupancies required a second level of ground-fault protection for the medical and dental offices. However, the second level of ground-fault protection was not required on the retail occupancies. This makes no sense if there were no critical procedures being performed in the doctor or dentist offices.

517.30(C)(3)

Essential Electrical Systems for Hospitals, Mechanical Protection of the Emergency System

NEC page 414

FIGURE 5.19

CHANGE TYPE: Revision (Proposal 15-42; Comment 15-25)

CHANGE SUMMARY: This section on protection of the emergency system wiring in hospitals was revised by eliminating the former general rule and reorganizing the five exceptions into positive language. The requirements are now in list format and allow listed flexible metal conduit and suitable metal-sheathed cable assemblies to be used where needed for fishing into existing walls and ceilings.

2005 CODE: "**(3) Mechanical Protection of the Emergency System.** The wiring of the emergency system of a hospital shall be mechanically protected ~~by installation in non-flexible metal raceways, or shall be wired with Type MI cable~~. Where installed as branch circuits in patient care areas, the installation shall comply with the requirements of 517.13(A) and (B). The following wiring methods shall be permitted:

(1) Non-flexible metal raceways, Type MI cable, or Schedule 80 rigid nonmetallic conduit. Nonmetallic raceways shall not be used for branch circuits that supply patient care areas.

(2) Where encased in not less than 50 mm (2 in.) of concrete, Schedule 40 rigid nonmetallic conduit flexible nonmetallic or jacketed metallic raceways, or jacketed metallic cable assemblies listed for installation in concrete. Nonmetallic raceways shall not be used for branch circuits that supply patient care areas.

Article 517 Health Care Facilities
Part III Essential Electrical System
517.30 Essential Electrical Systems for Hospitals
517.30(C) Wiring Requirements
517.30(C)(3) Mechanical Protection of the Emergency System

Permitted Wiring Methods

All nonflexible metal raceways	All Locations
Type MI cable	All Locations
Schedule 80 RNMC	Not permitted to contain branch circuits for patient care areas
Schedule 40 RNMC	In 2 inches of concrete, not permitted to contain branch circuits for patient care areas
Flexible nonmetallic conduit	In 2 inches of concrete, not permitted to contain branch circuits for patient care areas
Jacketed metallic raceways	In 2 inches of concrete
Jacketed metallic cable assemblies	Where the cable is listed for installation inconcrete, in 2 inches of concrete
Listed Flexible metal raceways	For use in listed headwalls, office furnishings, where fished into existing walls/ceilings, where necessary for flexible connection to equipment
Listed metal sheathed cable	For use in listed headwalls, office furnishings, where fished into existing walls/ceilings, where necessary for flexible connection to equipment
Flexible cord	Where necessary for connecting utilization equipment to the emergency system
Class 2/3	Secondary/signaling system cables

NOTE THAT 517.13 MANDATES ADDITIONAL GROUNDING REQUIREMENTS FOR PATIENT CARE AREAS

(3) Listed flexible metal raceways and listed metal-sheathed cable assemblies in any of the following:

(a) Where in listed prefabricated medical headwalls

(b) In listed office furnishings

(c) Where fished into existing walls or ceilings, not otherwise accessible and not subject to physical damage

(d) Where necessary for flexible connection to equipment

(4) Flexible power cords of appliances or other utilization equipment connected to the emergency system

(5) Secondary circuits of Class 2 or Class 3 communication or signaling systems

~~Exception No. 1: Flexible power cords of appliances, or other utilization equipment, connected to the emergency system shall not be required to be enclosed in raceways.~~

~~Exception No. 2: Secondary circuits of transformer-powered communications or signaling systems shall not be required to be enclosed in raceways unless otherwise specified by Chapters 7 or 8.~~

~~Exception No. 3: Schedule 80 rigid non-metallic conduit shall be permitted if the branch circuits do not serve patient care areas and it is not prohibited elsewhere in this Code.~~

~~Exception No. 4: Where encased in not less than 50 mm (2 in.) of concrete, Schedule 40 rigid nonmetallic conduit or electrical non-metallic tubing shall be permitted if the branch circuits do not serve patient care areas.~~

~~Exception No. 5: Flexible metal raceways and cable assemblies shall be permitted to be used in listed prefabricated medical headwalls, listed office furnishings, or where necessary for flexible connection to equipment.~~

FPN: See 517.13 for additional grounding requirements in patient care areas."

CHANGE SIGNIFICANCE: Several proposals and comments were made to revise this important section, whose purpose is to add a level of mechanical protection to help ensure that an emergency system will be functional when needed in hospitals. Significant improvements have been made to the readability and user friendliness of the section by reorganizing it into list format.

The most significant changes to the wiring methods permitted are in (C)(3)(3). The uses permitted for flexible metal conduit are clarified, and permission has been added to use "listed metal-sheathed cable assemblies" for protection of the emergency system in hospitals.

Where used as a wiring method in patient care areas, listed flexible metal conduit is required to contain an equipment grounding conductor and is limited in length to 6 feet, with contained circuit conductors having an overcurrent protection not exceeding 20 amperes. Refer to the discussion on the changes in 517.13 for additional information on the cable-type wiring methods permitted in patient care areas.

Article 518, 518.2

Assembly Occupancies

NEC page 425

CHANGE TYPE: Revision (Proposals 15-57 and 15-63; Comment 15-34)

CHANGE SUMMARY: The title of this article has changed from "Places of Assembly" to "Assembly Occupancies" to coordinate with other NFPA codes and standards. The examples of assembly occupancies have been revised and the application of Article 518 to multiple occupancies is now clearer.

2005 CODE: "ARTICLE 518 ~~Places of~~ Assembly <u>Occupancies</u>

518.1 Scope. <u>Except for the assembly occupancies explicitly covered by 520.1</u>, this article covers all buildings or portions of buildings or structures designed or intended for the <u>gathering together</u> ~~assembly~~ of 100 or more persons <u>for such purposes as deliberation, worship, entertainment, eating, drinking, amusement, awaiting transportation, or similar purposes.</u>

518.2 General Classifications.

(A) Examples. Places of assembly shall include but not be limited to the following:

Armories
Assembly halls
Auditoriums
~~Auditoriums within~~
    ~~Business establishments~~
    ~~Mercantile establishments~~
    ~~Other occupancies~~
    ~~Schools~~
Bowling lanes
~~Church chapels~~
Club rooms
Conference rooms

FIGURE 5.20

Courtrooms

Dance halls

Dining <u>and drinking</u> facilities

Exhibition halls

Gymnasiums

Mortuary chapels

Multipurpose rooms

Museums

Places of awaiting transportation

<u>Places of religious worship</u>

Pool rooms

Restaurants

Skating rinks

(B) Multiple Occupancies. <u>Where an assembly occupancy forms a portion of a building containing other occupancies, Article 518 applies only to that portion of the building considered an assembly occupancy.</u> Occupancy of any room or space for assembly purposes by less than 100 persons in a building of other occupancy, and incidental to such other occupancy, shall be classified as part of the other occupancy and subject to the provisions applicable thereto."

CHANGE SIGNIFICANCE: During the last Code cycle, Panel 15 voted to hold Comment 15-22 on Proposal 15-11. The chair of CMP-15 was directed by the Technical Correlating Committee to form a task group to review the comment, submitted during the 2002 Code cycle, and to formulate a proposal to clarify the wording of the scope of Article 518. A task group consisting of members of CMP-15, NFPA 101, and NFPA 5000 was formed and the matter was appropriately discussed and acted upon.

The task group decided that the definition of *assembly occupancy* should be consistent in the Building Code, NFPA 5000, the Life Safety Code, NFPA 101, and the NEC. The Fire Prevention Code, NFPA 1, NFPA 101, and NFPA 5000 define an assembly occupancy as described above in the Code excerpt. In addition, NFPA 1, NFPA 101, and NFPA 5000 make reference to NFPA 70 for the electrical aspects of a building. In order to effect proper cross-reference it is necessary to have consistent language in each of the Codes. However, the NEC remains unique among the NFPA family of Codes in that it continues to require compliance with Article 518 for assembly occupancies when 100 or more persons will occupy space, while other Codes in the NFPA family of Codes trigger at 50 or more persons.

518.4(C)

Assembly Occupancies, Wiring Methods, Spaces with Finish Rating

NEC page 426

CHANGE TYPE: Revision (Proposal 15-68)

CHANGE SUMMARY: The Panel voted to remove three types of occupancies that were previously permitted to be wired with electrical non-metallic tubing or rigid non-metallic conduit if enclosed in building construction or spaces that have at least a 15-minute finish rating.

2005 CODE: "**(C) Spaces with Finish Rating.** Electrical nonmetallic tubing and rigid nonmetallic conduit shall be permitted to be installed in club rooms, ~~college and university classrooms,~~ conference and meeting rooms in hotels or motels, courtrooms, ~~drinking establishments,~~ dining facilities, restaurants, mortuary chapels, museums, ~~passenger stations and terminals of air, surface, underground, and marine public transportation facilities,~~ libraries, and places of religious worship where the following apply:

(1) The electrical nonmetallic tubing or rigid non-metallic conduit is installed concealed within walls, floors, and ceilings where the walls, floors, and ceilings provide a thermal barrier of material that has at least a 15-minute finish rating as identified in listings of fire-rated assemblies.

(2) The electrical nonmetallic tubing or rigid non-metallic conduit is installed above suspended ceilings where the suspended ceilings provide a thermal barrier of material that has

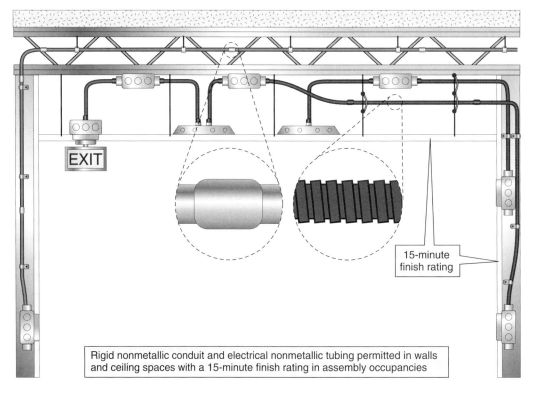

Rigid nonmetallic conduit and electrical nonmetallic tubing permitted in walls and ceiling spaces with a 15-minute finish rating in assembly occupancies

FIGURE 5.21

at least a 15-minute finish rating as identified in listings of fire-rated assemblies.

Electrical non-metallic tubing and rigid non-metallic conduit are not recognized for use in other space used for environmental air in accordance with 300.22(C).

FPN: A finish rating is established for assemblies containing combustible (wood) supports. The finish rating is defined as the time at which the wood stud or wood joist reaches an average temperature rise of 121°C (250°F) or an individual temperature rise of 163°C (325°F) as measured on the plane of the wood nearest the fire. A finish rating is not intended to represent a rating for a membrane ceiling."

CHANGE SIGNIFICANCE: The Code Panel voted by a 16 to 1 margin to accept this proposal to remove the three types of occupancies on the list that are permitted to be wired with electrical non-metallic tubing (ENT) or rigid non-metallic conduit (RNC). A voluminous substantiation was submitted to support the request. (The Panel's lone dissenter submitted a similarly extensive explanation of negative vote.) One comment from an affirmative voter explained, "Some of these changes are appropriate because they are not assembly occupancies or are not in the current form used earlier in the article."

525.10, 525.11

Services, Multiple Sources of Supply

NEC page 435

CHANGE TYPE: Revision (Proposals 15-91 and 15-92; Comment 15-62)

CHANGE SUMMARY: Text has been deleted from the previous 525.10 and 525.11, which referred to the entire text of Articles 445, 450, and 230. These changes were made to recognize 90.3, which provides that the rules of Chapters 1 through 4 of the Code apply unless amended in Chapters 5, 6, or 7. Other changes were made to require services or generators together when they serve rides or other structures that are less than 12 feet apart.

2005 CODE: "~~525.10 Separately Derived Systems.~~

~~(A) Generators. Generators shall comply with the requirements of Article 445.~~

~~(B) Transformers. Transformers shall comply with the applicable requirements of 240.4(A), (B)(3), and (C); 250.30; and Article 450.~~

525.10 ~~525.11~~ **Services.** Services shall ~~be installed in accordance with the applicable requirements of Article 230 and, in addition, shall~~ comply with 525.10(A) and 525.10(B) ~~525.11(A) and (B)~~.

(A) Guarding. Service equipment shall not be installed in a location that is accessible to unqualified persons, unless the equipment is lockable.

(B) Mounting and Location. Service equipment shall be mounted on a solid backing and be installed so as to be protected from the weather, unless of weather proof construction.

525.11 Multiple Sources of Supply. Where multiple services or separately derived systems or both supply rides, attractions, and other structures, all sources of supply that serve rides, attractions, or other structures separated by less than 3.7 m (12 ft.) shall be bonded to the same grounding electrode system."

FIGURE 5.22

CHANGE SIGNIFICANCE: Changes have been made to these sections to comply with the NEC style manual and to recognize the organization of the Code provided in 90.3. The rules in Chapters 1 through 4 of the Code apply generally unless they are amended, supplemented, or changed in some way in Chapters 5, 6, or 7. As a result, the general references to Articles 230, 445, and 450 have been deleted.

New requirements have been added to 525.11 to cover the situation where multiple sources of supply may be installed for a carnival, circus, fair, or similar event. Power for rides is often supplied from more than one generator or from a combination of generators and utility-supplied services. This new section provides that where multiple sources of supply have been installed and rides, attractions, or other structures are separated by less than 12 feet, they shall be bonded to the same grounding electrode system. This revision is an attempt to keep all conductive parts within reach of one or more persons at the same potential to reduce shock hazards.

525.23

Ground-Fault Circuit-Interrupter (GFCI) Protection

NEC page 436

CHANGE TYPE: Revision (Proposal 15-94)

CHANGE SUMMARY: This section on GFCI requirements for carnivals, circuses, fairs, and similar events has been changed extensively. It now includes where GFCI protection is required, where it is not required, and where it is not permitted.

2005 CODE: "**525.23 Ground-Fault Circuit-Interrupter (GFCI) Protection ~~for Personnel~~.**

(A) <u>Where GFCI Protection Is Required</u> ~~General-Use 15- and 20-Ampere, 125-Volt Receptacles~~. ~~All 125-volt, single-phase, 15- and 20-ampere receptacle outlets that are in use by personnel shall have listed ground-fault circuit-interrupter protection for personnel.~~ The ground-fault circuit interrupter shall be permitted to be an integral part of the attachment plug or located in the power-supply cord, within 300 mm (12 in.) of the attachment plug. ~~For the purposes of this section,~~ Listed cord sets incorporating ground-fault circuit-interrupter protection for personnel shall be permitted. ~~Egress lighting shall not be connected to the load-side terminals of a ground-fault circuit-interrupter receptacle.~~

(1) <u>125-volt, single-phase, 15- and 20-ampere non-locking-type receptacles used for disassembly and reassembly or readily accessible to the general public</u>

(2) <u>Equipment that is readily accessible to the general public and supplied from a 125-volt, single-phase, 15- or 20-ampere branch circuit</u>

(B) <u>Where GFCI Protection Is Not Required</u> ~~Appliance Receptacles~~. Receptacles <u>that only facilitate quick disconnecting and reconnecting of electrical equipment shall not be required to be provided with GFCI protection. These receptacles shall be of the locking type.</u>

FIGURE 5.23

~~supplying items, such as cooking and refrigeration equipment, that are incompatible with ground-fault circuit-interrupter devices shall not be required to have ground-fault circuit-interrupter protection.~~

(C) <u>Where GFCI Protection Is Not Permitted</u> ~~Other Receptacles~~. <u>Egress lighting shall not be protected by a GFCI.</u> ~~Other receptacle outlets not covered in 525.23(A) or (B) shall be permitted to have ground-fault circuit-interrupter protection for personnel, or a written procedure shall be continuously enforced at the site by one or more designated persons to ensure the safety of equipment grounding conductors for all cord sets and receptacles, as described in 527.6(B)(2).~~"

CHANGE SIGNIFICANCE: This proposal was originally a comment on a proposal for the 2002 Code that was held over for the 2005 Code. As can be seen, the requirements have been significantly changed.

GFCI protection is required in general areas that are accessible to the public, as are non-locking-type receptacles used for disassembly and reassembly of equipment. Receptacles that are of the locking type and only facilitate the quick disconnecting and reconnecting of electrical equipment are not required to have GFCI protection. The (receptacles') power supply to egress lighting is not permitted to have GFCI protection.

547.5(G)

Agricultural Buildings, Receptacles

NEC page 444

CHANGE TYPE: Revision (Proposal 19-10a; Comment 19-10)

CHANGE SUMMARY: Changes have been made to locate all of the requirements for ground-fault circuit-interrupter (GFCI) protection in one location.

2005 CODE: "**(G) Receptacles.** All 125-volt, single-phase, 15- and 20-ampere general-purpose receptacles installed in the following locations shall have ground-fault circuit-interrupter protection for personnel:

(1) Areas having an equipotential plane

(2) Outdoors

(3) Damp or wet locations

(4) Dirt confinement areas for livestock"

CHANGE SIGNIFICANCE: CMP-19 began this series of proposals and comments by creating a Panel proposal that included the requirement "Other circuits providing electric power to metallic equipment that may become energized and is accessible to livestock in dirt confinement areas as covered in 547.10(B) shall have ground-fault protection of equipment." This equipment should not be confused with ground-fault circuit-interrupters that provide protection against electric shock for personnel. Ground-fault protection of equipment will trip at about 30 mA of leakage current to ground. This is about six times the trip level of the traditional GFCI.

Changes were made at the comment stage to delete the requirements for ground-fault protection of equipment and add item (4) to this section. This requirement was previously located in 547.10(B).

FIGURE 5.24

CHANGE TYPE: Revision (Proposal 19-12a; Comments 19-16, 19-17, and 19-18)

CHANGE SUMMARY: One of the key revisions to this section is that a site-isolating device, as now defined in 547.2, is required to be pole-mounted. The section has been reorganized in compliance with the NEC style manual.

2005 CODE: "**547.9 Electrical Supply to Building(s) or Structure(s) From a Distribution Point.**
(A) Site-Isolating Device

(1) Where required

(2) Location

(3) Operation

(4) Bonding provisions

(5) Grounding

(6) Rating

(7) Overcurrent protection

(8) Accessibility

(9) Series devices

(B) Service Disconnecting Means and Overcurrent Protection at the Building(s) or Structure(s)

(1) Conductor sizing

(2) Conductor installation

(3) Grounding and bonding

    (a) System with grounded neutral conductor

    (b) System with separate equipment grounding conductor

547.9

Electric Supply to Building(s) or Structure(s) From a Distribution Point

NEC pages 444, 445

FIGURE 5.25

(C) Service Disconnecting Means and Overcurrent Protection at the Distribution Point
(D) Direct-Buried Equipment Grounding Conductors"

CHANGE SIGNIFICANCE: Where a site-isolating device is readily accessible at grade level, it must meet all the provisions for service equipment, including overcurrent protection and short-circuit current ratings. This revision to 547.9 takes these provisions into account and accomplishes several objectives. First, the section has been revised to comply with the NEC Style Manual. Second, the proposal incorporates in principle or in part the technical changes in the recommendations of Proposals 19-14, 19-16, 19-17, 19-18, 19-19, 19-21, and 19-23. Third, the panel has revised existing text covering grounding at and accessibility to the site-isolating device.

This overall proposal will clarify the requirements for farm premises distribution.

547.10

Equipotential Plane Requirements for Agricultural Buildings, Equipotential Planes and Bonding of Equipotential Planes
NEC page 445

CHANGE TYPE: Revised (Proposals 19-10a and 19-23a; Comment 19-25)

CHANGE SUMMARY: The requirements for providing an equipotential plane to reduce or control tingle or stray voltages in animal confinement have been revised. The previous language on where equipotential planes are not required has been deleted.

2005 CODE: "547.10 Equipotential Planes and Bonding of Equipotential Planes. <u>The installation and bonding of equipotential planes shall comply with 547.10(A) and 547.10(B).</u> For the purposes of this section, the term *livestock* shall not include poultry.

(A) <u>Where Required</u> ~~Areas Requiring Equipotential Planes~~. Equipotential planes shall be installed in all concrete floor confinement areas <u>in</u> ~~of~~ livestock buildings, <u>and in all outdoor confinement areas such as feedlots, containing metallic equipment that may become energized and is accessible to livestock</u>. ~~that contain metallic equipment that is accessible to animals and likely to become energized. Outdoor confinement areas, such as feedlots, shall have equipotential planes installed around metallic equipment that is accessible to animals and likely to become energized.~~ The equipotential plane shall encompass the area <u>where the livestock</u> ~~around the equipment where the animal~~ stands while accessing <u>metallic</u> ~~the~~ equipment <u>that may become energized</u>.

(B) ~~Areas Not Requiring Equipotential Planes. Equipotential planes shall not be required in dirt confinement areas containing metallic equipment that is accessible to animals and likely to become energized. All circuits providing electric power to equipment that is accessible to animals in dirt confinement areas shall have GFCI protection.~~

FIGURE 5.26

(B) Bonding. Equipotential planes shall be bonded to the electrical grounding system. The bonding conductor shall be copper, insulated, covered or bare, and not smaller than 8 AWG. The means of bonding to wire mesh or conductive elements shall be by pressure connectors or clamps of brass, copper, copper alloy, or an equally substantial approved means. Slatted floors that are supported by structures that are a part of an equipotential plane shall not require bonding.

FPN NO. 1: Methods to establish equipotential planes are described in American Society of Agricultural Engineers (ASAE) EP473-2001, *Equipotential Planes in Animal Containment Areas.*

FPN NO. 2: Low grounding electrode system resistances may reduce potential differences in livestock facilities."

CHANGE SIGNIFICANCE: Changes to this section emphasize that an equipotential plane in animal confinement areas is required where the livestock are exposed to accessible metallic parts that may become energized. The previous language used the phrase "likely to become energized."

The requirements continue to apply to both agricultural buildings with concrete floors and outdoor confinement areas. The term *animal* has been changed to *livestock* throughout the section.

CHANGE TYPE: Revision (Proposals 19-142 and 14-110)

CHANGE SUMMARY: Section 551.21 now covers motor fuel dispensing stations at marinas and boatyards. Changes require that the installation of electrical wiring for power and lighting be on the opposite side of piers from liquid piping systems. A new FPN refers to NFPA 30A and NFPA 303. Also, another change requires electrical equipment located in facilities for the repair of marine craft to comply with the rules in Article 511, in addition to the requirements of this article.

2005 CODE: "**555.21** Motor Fuel Gasoline **Dispensing Stations— Hazardous (Classified) Locations.** Electrical wiring and equipment located at or serving motor fuel gasoline dispensing stations shall comply with Article 514 in addition to the requirements of this article. All electrical wiring for power and lighting shall be installed on the side of the wharf, pier, or dock opposite from the liquid piping system.

FPN: For additional information, see NFPA 303-2000, *Fire Protection Standard for Marinas and Boatyards*, and NFPA 30A-2003, *Motor Fuel Dispensing Facilities and Repair Garages.*

555.22 Repair Facilities—Hazardous (Classified) Locations. Electrical wiring and equipment located at facilities for the repair of marine craft containing flammable or combustible liquids or gases shall comply with Article 511 in addition to the requirements of this article."

555.21 and 555.22

Motor Fuel Dispensing Stations—Hazardous (Classified) Locations, Repair Facilities— Hazardous (Classified) Locations
NEC page 480

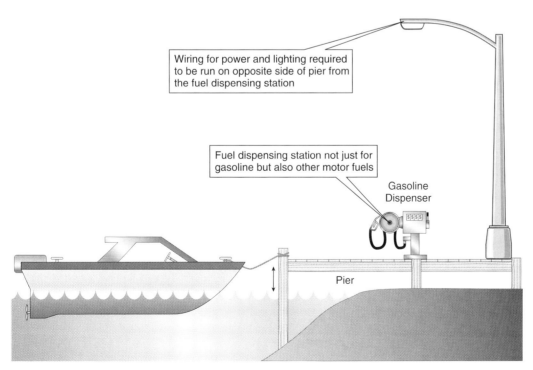

Wiring for power and lighting required to be run on opposite side of pier from the fuel dispensing station

Fuel dispensing station not just for gasoline but also other motor fuels

Gasoline Dispenser

Pier

FIGURE 5.27

CHANGE SIGNIFICANCE: The 555.21 title change to "motor fuel" dispensing stations reflects the change in terminology in Article 514 and NFPA 30A. (This term is also used in the first sentence of 555.21.) The second sentence is added to reflect an existing requirement in NFPA 30A 11.5.4. Since this is an installation requirement for the wiring system, it is felt that this requirement should be stated here in the NEC rather than have to apply a reference to another document that may not be readily available to the installing contractor. Also, the FPN has been added back into this section after it was inadvertently deleted in the 2002 rewrite.

The panel rejected the concept of 555.22 during the 1996 Code cycle (Proposals 14-201 and 202) because of a lack of subject coverage in NFPA 88A and 88B. That position left a vacuum in NFPA standards on this topic. The rewritten NFPA 303 does not address this topic unless the facility is part of a marina, and even in that case, Chapter 3 of that standard generally defers to the NEC with respect to hazardous (classified) locations. Adding a requirement for the repair of water or marine craft strengthens the appropriate rules in place for motorboat repair facilities. Occasional leakage of gasoline in these boats or vessels is a major concern, as is the extensive use of flexible cords in areas with moist concrete floors.

CHANGE TYPE: Relocation (Proposal 3-108)

CHANGE SUMMARY: Article 527, which addresses temporary installations, has been relocated to Article 590.

2005 CODE: **"Article 590 Temporary Installations**
590.1 Scope. The provisions of this article apply to temporary electrical power and lighting installations."

CHANGE SIGNIFICANCE: This article was Article 305 for many Code cycles, with a title of "Temporary Wiring." It was relocated to Article 527 during the reorganization of Chapter 3 during the 2002 NEC cycle since it was felt that the material did not fit well in Chapter 3.

The entertainment industry then asked for the article on temporary wiring to be moved away from "their articles" 520, 530, and 540, which have been in place for over 50 years, with 518 added more than 25 years ago. Article 525, the newest "entertainment industry article," was added in 1996.

As of this Code cycle, this text has been relocated to Article 590.

Article 590

Temporary Installations
NEC page 480

FIGURE 5.28

590.4(B) and (C)

Temporary Installations, Feeders and Branch Circuits

NEC page 481

CHANGE TYPE: Revision (Proposal 3-112)

CHANGE SUMMARY: The Code Panel has accepted the language from a Tentative Interim Amendment that was added to the 2002 NEC as No. 02-1. As seen in this revision, it clarifies that Type NM cable may be used for temporary installations in buildings of any height or construction type.

2005 CODE: "**(B) Feeders.** Overcurrent protection shall be provided in accordance with 240.4, 240.5, 240.100, and 240.101. Feeders shall be protected as provided in Article 240. They shall originate in an approved distribution center. Conductors shall be permitted within cable assemblies or within multiconductor cords or cables of a type identified in Table 400.4 for hard usage or extra-hard usage. For the purpose of this section, Type NM and Type NMC cables shall be permitted to be used in any dwelling, building, or structure without any height limitation or limitation by building construction type and without concealment in walls, floors, or ceilings.

Exception: Single insulated conductors shall be permitted where installed for the purpose(s) specified in 590.3(C), where accessible only to qualified persons.

(C) Branch Circuits. All branch circuits shall originate in an approved power outlet or panelboard. Conductors shall be permitted within cable assemblies or within multiconductor cord or cable of a type identified in Table 400.4 for hard usage or extra-hard usage. All Conductors shall be protected from overcurrent as provided in Article 240.4, 240.5, and 240.100. For the purposes of this section, Type NM

FIGURE 5.29

and Type NMC cables shall be permitted to be used in any dwelling, building, or structure without any height limitation <u>or limitation by building construction type and without concealment in walls, floors, or ceilings</u>.

Exception: Branch circuits installed for the purposes specified in 590.3(B) or 590.3(C) shall be permitted to be run as single insulated conductors. Where the wiring is installed in accordance with 590.3(B), the voltage to ground shall not exceed 150 volts, the wiring shall not be subject to physical damage, and the conductors shall be supported on insulators at intervals of not more than 3.0 m (10 ft); or, for festoon lighting, the conductors shall be arranged so that excessive strain is not transmitted to the lampholders.

CHANGE SIGNIFICANCE: Previously, the language of 334.10 in the 2002 NEC could be interpreted to limit the use of Type NM or NMC cables for temporary installations in some construction types unless the cables were protected by a 15-minute fire-rated finish. That was never the intent of CMP-3, which has jurisdiction over Article 590.

CHAPTER 6

Special Equipment, Articles 600–695

600.2, 600.8, 600.24

Section Sign, Enclosures, Class 2 Power Sources

NEC pages 483, 484, 486

CHANGE TYPE: New (Proposal 18-107 and 18-110)

CHANGE SUMMARY: A new definition of *Section Sign* has been added to 600.2. Live parts, other than lamps and neon tubing, are now required to be enclosed. Requirements for signs supplied by Class 2 power supplies have also been added.

2005 CODE: "**600.2 Section Sign.** A sign or online lighting system, shipped as subassemblies, that requires field-installed wiring between the subassemblies to complete the overall sign."

"**600.8 Enclosures.** Live parts, other than lamps and neon tubing shall be enclosed. Transformers and power supplies provided with an internal enclosure, including a primary and secondary circuit splice enclosure, shall not require an additional enclosure.

~~Exception: A transformer or electronic power supply provided with an integral enclosure, including a primary and secondary circuit splice enclosure, shall not be required to be provided with an additional enclosure.~~

(A) Strength. Enclosures shall have ample structural strength and rigidity.

(B) Material. Sign and outline lighting system enclosures shall be constructed of metal or shall be listed.

(C) Minimum Thickness of Enclosure Metal. Sheet copper or aluminum shall be at least 0.51 mm (0.020 in.) thick. Sheet steel shall be at least 0.41 mm (0.016 in.) thick.

(D) Protection of Metal. Metal parts of equipment shall be protected from corrosion."

"**600.24 Class 2 Power Sources.** In addition to the requirements of Article 600, signs and online lighting systems supplied by Class 2 transformers, power supplies, and power sources shall comply with 725.41."

FIGURE 6.1

CHANGE SIGNIFICANCE: Section signs are subassemblies of an over-all sign shipped from the manufacturer's plant and assembled on-site. Field-installed wiring is required between sign sections. Since these signs are required to be listed, the manufacturer is required to include installation instructions with the sign sections. Section signs are often individually lettered signs and are shipped from the factory with indi-vidual listing marks such as "Sign Unit 1 of 6," "Sign Unit 2 of 6," etc.

The term *Section Signs* has not previously been used in the NEC. New 600.12 requires that "field-installed secondary wiring of section signs shall comply with 600.31 if 1000 volts or less, or with 600.32 if over 1000 volts."

The change to 600.8 moves the recognition for an enclosure for a sign transformer or power supply from being an exception to the main rule itself. This section recognizes that the enclosure for the transformer is adequate to enclose both the primary and secondary conductors of the sign transformer or power supply.

600.24 adds a requirement that signs and outline lighting systems supplied by Class 2 transformers must comply with 725.41. These signs may not be available listed and consist of several parts that may be assembled on-site. Some signs have light emitting diodes (LEDs). Sec-tion 725.41 provides requirements for Class 2 and Class 3 power sup-plies where transformers and power supplies are required to be listed.

604.6(E)

Manufactured Wiring Systems, Securing and Supporting

NEC page 488

CHANGE TYPE: Revision (Proposal 19-155a)

CHANGE SUMMARY: The title and text of the section has been changed to include both securing and supporting the manufactured wiring in accordance with the applicable cable or conduit article.

2005 CODE: "(E) Securing and Supporting. Manufactured wiring systems shall be <u>secured and</u> supported in accordance with the applicable cable or conduit article for the cable or conduit type employed."

CHANGE SIGNIFICANCE: Changes to this section further emphasize the link between the manufactured wiring systems in Article 604 and the parent article for the cable or raceway used to manufacture the wiring system. The following table shows the parent article for the wiring methods permitted to be used and the section requiring supporting and securing the wiring method.

Wiring Method	Securing and Supporting Rule	General Rule*
Type AC cable	320.30	Every 4 $^{1}/_{2}$ ft and within 12 in. of boxes
Type MC cable	330.30	Every 6 ft and within 12 in. of boxes
Flexible metal conduit	348.30	Every 4 $^{1}/_{2}$ ft and within 12 in. of boxes
Liquidtight flexible metal conduit	350.30	Every 4 $^{1}/_{2}$ ft and within 12 in. of boxes
Other cables (identified in Section ...)	725.61, 800.113, 820.113, 830.179	See respective article

* See exceptions where wiring method is permitted to be unsupported.

FIGURE 6.2

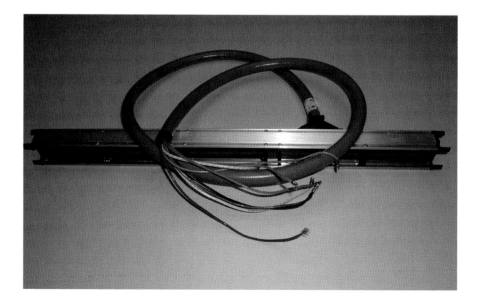

605.6, 605.7

Office Furnishings, Fixed-Type Partitions and Freestanding-Type Partitions
NEC page 489

CHANGE TYPE: Revision (Proposals 18-120a and 18-121)

CHANGE SUMMARY: For both fixed-type partitions in 605.6 and freestanding-type partitions in 605.7, multiwire branch circuits supplying the office partitions are required to be supplied with a simultaneous disconnecting means at the panelboard where the branch circuit originates.

2005 CODE: "**605.6 Fixed-Type Partitions.** Wired partitions that are fixed (secured to building surfaces) shall be permanently connected to the building electrical system by one of the wiring methods of Chapter 3. <u>Multiwire branch circuits supplying power to the partition shall be provided with a means to disconnect simultaneously all ungrounded conductors at the panelboard where the branch circuit originates.</u>

"**605.7 Freestanding-Type Partitions.** Partitions of the freestanding type (not fixed) shall be permitted to be connected to the building electrical system by one of the wiring methods of Chapter 3. <u>Multiwire branch circuits supplying power to the permanently connected freestanding partitions shall be provided with a means to disconnect simultaneously all ungrounded conductors at the panelboard where the branch circuit originates.</u>"

CHANGE SIGNIFICANCE: These two types of equipment are often supplied with circuits from different panelboards. These circuits are present in the same junction box used to feed this partition furniture. The FPN at the end of Article 605 refers to 210.4 for circuits supplying partitions in 605.6 and 605.7, but this is really not an enforceable part of the Code article. Often, the architect or engineer calls for the installer

FIGURE 6.3

to use specific circuits for the supply to the office partition, and these branch circuits may be from different panelboards.

Maintenance staff is not usually familiar with the source of circuits; when they get the trouble call for a problem with the partition furniture they often wind up getting hurt. Because these types of partitions are moved frequently, the electrician/contractor will be asked to disconnect and reconnect the furniture. Again, those individuals not familiar with the original install will get hurt. But by grouping the circuits together in the panelboard and by terminating the conductors on a multipole circuit breaker, this hazard may be reduced.

Multiwire branch circuits are not permitted to be used for individual or groups of interconnected individual office partitions. See 605.8(D).

CHANGE TYPE: Revision (Proposal 12-15; Comment 12-10)

CHANGE SUMMARY: A bonding jumper is now required for trolley-frame and bridge-frame cranes. Bridge and trolley wheels and their respective tracks are no longer considered to provide effective grounding and bonding.

2005 CODE: "**610.61 Grounding.** All exposed non-current-carrying metal parts of cranes, monorail hoists, hoists, and accessories, including pendant controls, shall be metallically joined together into a continuous electrical conductor so that the entire crane or hoist will be grounded in accordance with Article 250. Moving parts, other than removable accessories, or attachments, that have metal-to-metal bearing surfaces shall be considered to be electrically connected to each other through bearing surfaces for grounding purposes. The trolley frame and bridge frame shall <u>not</u> be considered as electrically grounded through the bridge and trolley wheels and their respective tracks<u>. unless local conditions, such as paint or other insulating material, prevent reliable metal-to-metal contact. In this case.</u> A separate bonding conductor shall be provided."

CHANGE SIGNIFICANCE: The Code Panel understood the previous text to recognize the ever-present possibility for paint or other insulating materials to prevent reliable metal-to-metal contact between bridge and trolley wheels and their respective tracks. This indicates the need for a requirement for a separate bonding conductor to be installed if reliable metal-to-metal contact is not assured. The submitter of Proposal 12-15, an employee in the crane and hoist industry, proposed that the only way to assure this necessary bonding is by requiring the use of a separate grounding conductor in lieu of reliance on metal-to-metal contact between the bridge and trolley wheels and their respective tracks. This revision reflects this requirement.

610.61
Cranes and Hoists, Grounding
NEC page 494

FIGURE 6.4

620.22(A)

Elevators, Etc., Car Light Source

NEC page 499

CHANGE TYPE: Revision (Proposal 12-23)

CHANGE SUMMARY: A new paragraph has been added to this section to mandate that required lighting for an elevator car should not be connected to the load side of a ground-fault circuit-interrupter (GFCI).

2005 CODE: "**(A) Car Light Source.** A separate branch circuit shall supply the car lights, receptacle(s), auxiliary lighting power source, and ventilation on each elevator car. The overcurrent device protecting the branch circuit shall be located in the elevator machine room or control room/machinery space or control space.

Required lighting shall not be connected to the load side of a ground-fault circuit-interrupter."

CHANGE SIGNIFICANCE: The individual who submitted this proposal indicated that he has inspected numerous elevator installations and modifications and has seen car and car-top lighting circuits being connected to the load side of the required GFCI receptacle installed on top of the car. The Code Panel recognized that this should not be allowed for the obvious reason that if a portable tool being used on the car top trips the GFCI, all lighting in and on the car will be extinguished.

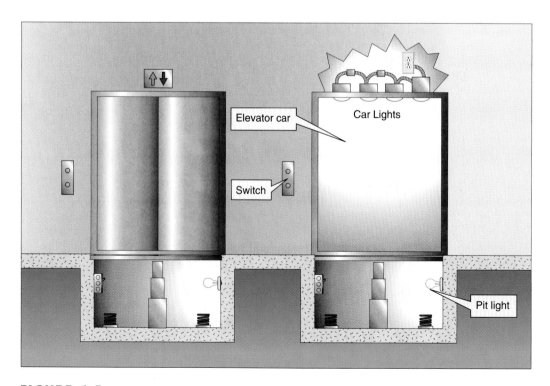

FIGURE 6.5

CHANGE TYPE: Revision (Proposal 12-33)

CHANGE SUMMARY: This revision adds "neighborhood electric vehicle" as a new category of electric vehicle. The term "neighborhood electric vehicle" is not further defined but refers to golf-cart-style vehicles or vehicles that are "street legal."

2005 CODE: "**625.2 Electric Vehicle.** An automotive-type vehicle for highway use, such as passenger automobiles, buses, trucks, vans, <u>neighborhood electric vehicles,</u> and the like, primarily powered by an electric motor that draws current from a rechargeable storage battery, fuel cell, photovoltaic array, or other source of electric current. For the purpose of this article, electric motorcycles and similar type vehicles and off-road self-propelled electric vehicles, such as industrial trucks, hoists, lifts, transports, golf carts, airline ground support equipment, tractors, boats, and the like, are not included."

CHANGE SIGNIFICANCE: Neighborhood electric vehicles are required to have automotive grade headlights, seat belts, windshields, brakes, and other safety equipment. Top speed is about 25 MPH. These vehicles are intended for use in retirement or other planned communities for short shopping trips and social and recreational purposes.

625.2
Electric Vehicle
NEC page 504

FIGURE 6.6

625.25, 625.26

Loss of Primary Source and Interactive Systems

NEC page 506

CHANGE TYPE: Revised and New (Proposals 12-34 and 12-35)

CHANGE SUMMARY: These changes recognize that new classes of hybrid electric vehicles are being developed with the capability to use an on-board generation system as either a standby power source or as a source of power production operating in parallel with a primary source of electricity.

2005 CODE: "**625.25 Loss of Primary Source.** Means shall be provided such that, upon loss of voltage from the utility or other electric system(s), energy cannot be back fed through the electric vehicle <u>and</u> the supply equipment to the premises wiring system <u>unless permitted by 625.26.</u> The electric vehicle shall not be permitted to serve as a standby power supply.

625.26 Interactive Systems. Electric vehicle supply equipment and other parts of the system, either on-board or off-board the vehicle, that are identified for and intended to be interconnected to a vehicle and also serve as an optional standby system or an electric power production source or provide for bi directional power feed shall be listed as suitable for that purpose. When used as an optional standby system, the requirements of Article 702 shall apply, and when used as an electric power production source, the requirements of Article 705 shall apply."

CHANGE SIGNIFICANCE: These changes allow hybrid electric vehicles to serve in the described capacity provided that they are listed for the purpose of a standby power supply. Where used to supply premises wiring, the installation must be in compliance with Article 702, Optional Standby Systems. Where used to back-feed into the utility electrical system, the installation must be in compliance with Article 705, Interconnected Electrical Power Production Sources.

FIGURE 6.7

CHANGE TYPE: Revision (Proposal 12-47; Comment 12-29)

CHANGE SUMMARY: Changes have been made to this section to correct the described application and remove language that, if interpreted literally, would prevent some acceptable support methods from being used.

2005 CODE: "**640.6 Mechanical Execution of Work.** Equipment and cables ~~cabling~~ shall be installed in a neat and workmanlike manner. Cables installed exposed on the surface of ceiling and sidewalls shall be supported ~~by the structural components of the building structure~~ in such a manner that the cables will not be damaged by normal building use. Such cables shall be supported ~~attached to structural components~~ by straps, staples, hangers, or similar fittings designed and installed so as not to damage the cable. The installation shall ~~also~~ conform to ~~with~~ 300.4(D) and 300.11."

CHANGE SIGNIFICANCE: Section 640.6 was revised during the 2002 Code cycle to remove the Fine Print Note referencing the ANSI/EIA/TIA cabling, wiring, and pathways standards. In so doing, the Panel included additional mechanical considerations in the body of the rule. Taken literally, all cables would have to be supported by building "structural components." This would preclude the attachment of cables to baseboards and walls.

This proposed change clarifies the intent of the rule, yet continues to advise the reader regarding the proper support of cables. It also makes "ceiling" plural. This is a companion proposal and is intended to correlate with similar proposals for 800.6, 820.6, 830.7, 725.6, 760.6, and 770.8.

640.6

Mechanical Execution of Work for Audio Signal Processing Equipment
NEC page 511

FIGURE 6.8

FPN to 670.1

Industrial Machinery, Scope

NEC page 525

CHANGE TYPE: Revision (Proposal 12-84)

CHANGE SUMMARY: This revision deletes section 670.5 and revises the FPN to 670.1 to direct the user of the Code to 110.26 and to NFPA 79 for working space rules.

2005 CODE: "**670.1 Scope.** This article covers the definition of, the nameplate data for, and the size and overcurrent protection of supply conductors to industrial machinery.

FPN: For information on the workspace requirements for equipment containing supply conductor terminals, see 110.26. For information on the workspace requirements for machine power and control equipment, see NFPA 79-2002, *Electrical Standard for Industrial Machinery.*"

"~~670.5 Clearance.~~ ~~Where the conditions of maintenance and supervision ensure that only qualified persons service the installation, the dimensions of the working space in the direction of access to live parts operating at not over 150 volts line to line or line to ground that are likely to require examination, adjustment, servicing, or maintenance while energized shall be a minimum of 750 mm (2½ ft). Where controls are enclosed in cabinets, the door(s) shall open at least 90 degrees or be removable.~~

~~*Exception: Where the enclosure requires a tool to open, and where only diagnostic and troubleshooting testing is involved on live parts, the clearances shall be permitted to be less than 750 mm (2½ ft).*~~"

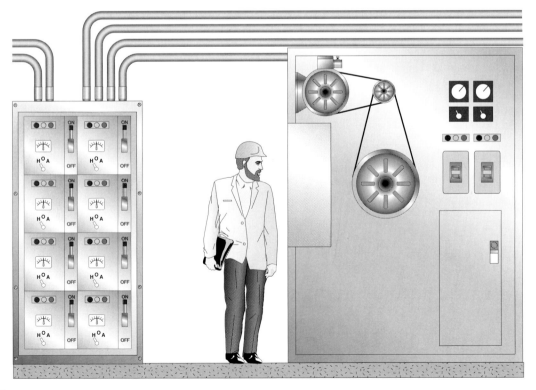

FIGURE 6.9

CHANGE SIGNIFICANCE: It was concluded that the intent of Section 670.5 was to allow different spacings than those in 110.26 due to the nuance of the machinery, as well as to compensate for the fact that NFPA 79 did not contain workspace requirements at the time. However, the 2002 edition of NFPA 79 now contains these requirements for equipment within its scope. Deleting 670.5 returns all workspace and access requirements to Section 110.26 due to the organization of the NEC as provided in 90.3.

The Task Group also concluded revision of the FPN to 670.1 would clarify the appropriate workspace requirements and guide the user to the appropriate standard.

680.8

Swimming Pools and Similar, Overhead Conductor Clearances

NEC page 530

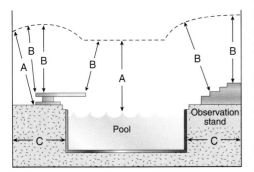

FIGURE 6.10

CHANGE TYPE: New (Proposal 17-61; Comment 17-102)

CHANGE SUMMARY: Two new sentences have been added to the beginning of this section to clarify the application of the measurements of overhead conductors as illustrated in Figure 680.8.

2005 CODE: "**680.8 Overhead Conductor Clearances.** <u>Overhead conductors shall meet the clearance requirements in this section. Where a minimum clearance from the water level is given, the measurement shall be taken from the maximum water level of the specified body of water.</u>

(A) Power. With respect to service drop conductors and open overhead wiring, swimming pool and similar installations shall comply with the minimum clearances given in Table 680.8 and illustrated in Figure 680.8.

FPN: Open overhead wiring as used in this article typically refers to conductor(s) not in an enclosed raceway."

CHANGE SIGNIFICANCE: These new sentences coordinate with the new term *body of water* introduced in the 2002 NEC and apply to any body of water to which Article 680 applies.

A new definition was added to the 2002 NEC to provide clarification for the height requirements for the deck boxes covered in 680.24. This revision would clarify the requirement for 680.8, which requires a minimum clearance from the water level as well as from the deck or other features.

The change to this section relies on the definition of *maximum water level*, which was added to the 2002 NEC: "The highest level that water can reach before it spills out." So far as Figure 680.8 is concerned, it can be argued that the maximum water level is the pool deck, since that is the highest level the water can reach. Though the pool is usually filled to some level below the deck level, the pool water level could reach the level of the deck during a rainy period or if the pool is overfilled. This maximum water level then would have dimension "A," shown to be from the deck level rather than from the normal water level.

CHANGE TYPE: Revision (Proposal 17-71)

CHANGE SUMMARY: This clarification of the rules on maintenance disconnecting means requires the means to be readily accessible.

2005 CODE: "**680.12 Maintenance Disconnecting Means.** One or more means to disconnect all ungrounded conductors shall be provided for all utilization equipment other than lighting. Each means shall be <u>readily</u> accessible and within sight from its equipment."

CHANGE SIGNIFICANCE: The disconnecting means for pool utilization equipment is now required to be "readily accessible" rather than just "accessible." *Readily accessible* is defined in Article 100 as "Capable of being reached quickly for operation, renewal, or inspections without requiring those to whom ready access is requisite to climb over or remove obstacles or to resort to portable ladders, and so forth." The previous term, *accessible*, is a lower level requirement.

680.12

Swimming Pools and Similar, Maintenance Disconnecting Means
NEC page 531

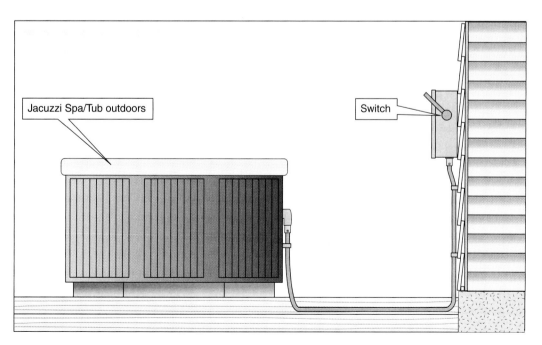

FIGURE 6.11

680.21 (A)(4)

Swimming Pools and Similar, One-Family Dwellings
NEC page 532

CHANGE TYPE: Revision (Proposal 17-75)

CHANGE SUMMARY: The sentence "Where run in a raceway, the equipment grounding conductor shall be insulated" has been deleted.

2005 CODE: "**(4) One-Family Dwellings.** In the interior of one-family dwellings, or in the interior of accessory buildings associated with a one-family dwelling, any of the wiring methods recognized in Chapter 3 of this *Code* ~~shall be permitted~~ that comply with the provisions of this paragraph <u>shall be permitted</u>. ~~Where run in a raceway, the equipment grounding conductor shall be insulated.~~ Where run in a cable assembly, the equipment grounding conductor shall be permitted to be uninsulated, but it shall be enclosed within the outer sheath of the cable assembly."

CHANGE SIGNIFICANCE: This revision applies to the wiring inside a one-family dwelling and coordinates with the permission to have uninsulated equipment grounding conductors in the wiring method, such as Type NM. It should be noted that this permission to use a cable wiring method applies only:

- To branch circuits for permanently installed pools
- To branch circuits for motors associated with permanently installed pools
- Inside a one-family dwelling
- Inside an accessory building associated with a one-family dwelling

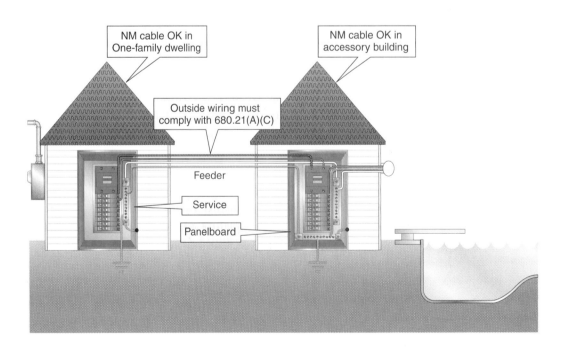

FIGURE 6.12

Wiring for motors outside a one-family dwelling, including between the dwelling and an associated accessory building, must be one of the wiring methods in 680.21(A)(1). Required wiring methods include rigid metal conduit, intermediate metal conduit, rigid non-metallic conduit, or Type MC cable listed for the location, such as PVC jacketed cable.

680.23 (B)(2)

Swimming Pools and Similar, Wiring Extending Directly to the Forming Shell

NEC page 533

CHANGE TYPE: Revision (Proposal 17-90)

CHANGE SUMMARY: This replaces the previously incorrect "equipment grounding conductor" with "bonding jumper" for the 8 AWG bonding conductor.

2005 CODE: "**(2) Wiring Extending Directly to the Forming Shell.** Conduit shall be installed from the forming shell to a ~~suitable~~ junction box or other enclosure <u>conforming to the requirements</u> ~~located as provided~~ in 680.24. Conduit shall be rigid metal, intermediate metal, liquidtight flexible nonmetallic, or rigid nonmetallic.

(a) **Metal Conduit.** Metal conduit shall be approved and shall be of brass or other approved corrosion-resistant metal.

(b) **Nonmetallic Conduit.** Where a nonmetallic conduit is used, an 8 AWG insulated solid or stranded copper <u>bonding jumper</u> ~~equipment grounding conductor~~ shall be installed in this conduit unless a listed low-voltage lighting system not requiring grounding is used. The <u>bonding jumper</u> ~~equipment grounding conductor~~ shall be terminated in the forming shell, junction box or transformer enclosure, or ground-fault circuit-interrupter enclosure. The termination of the 8 AWG <u>bonding jumper</u> ~~equipment grounding conductor~~ in the forming shell shall be covered with, or encapsulated in, a listed potting compound to protect the connection from the possible deteriorating effect of pool water."

CHANGE SIGNIFICANCE: These changes primarily correct the use of terms used to identify the conductor required to be installed between the junction box for a pool underwater light and the forming shell for a wet-niche light. Previously, the conductor was improperly described as an "equipment grounding conductor," rather than a "bonding jumper." The bonding jumper is used to equalize any voltage in or around the pool rather than to carry current during a line-to-ground fault as is the purpose of an equipment grounding conductor.

Identical changes were made to Section 680.27(A)(2).

FIGURE 6.13

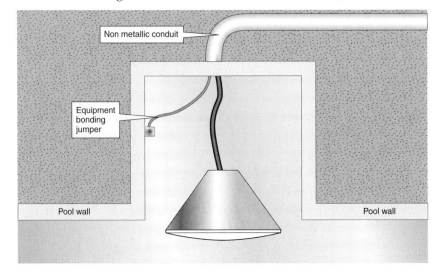

CHANGE TYPE: New (Proposal 17-98; Comment 17-139)

CHANGE SUMMARY: This revision adds a new Subsection (6) on servicing underwater luminaires.

2005 CODE: "**(6) Servicing.** All luminaires shall be removable from the water for relamping or normal maintenance. Luminaires shall be installed in such a manner that personnel can reach the luminaire for relamping, maintenance, or inspection while on the deck or equivalently dry location."

CHANGE SIGNIFICANCE: Swimming pools are now being installed in such a manner that the adjacent structural foundation for the building is part of the pool wall. This is fairly common at higher end residences. When an underwater pool light is installed on the common wall and there is no deck above the fixture, the fixture cannot be reached for servicing unless a person is in the water or the pool is partially drained. Even then people may still be standing in the water for relamping. This is a real potential shock hazard, especially if someone inside the residence does not realize there is someone working on the pool lighting and accidentally energizes the luminaire.

A similar requirement has been included for some time in 680.51(F) for servicing underwater luminaires (fixtures).

680.23 (B)(6)
Wet-Niche Luminaires (Fixtures), Servicing
NEC page 534

FIGURE 6.14

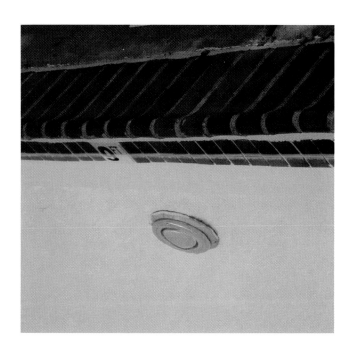

680.26

Swimming Pools and Similar, Equipotential Bonding

NEC page 536

CHANGE TYPE: Revision (Proposal 17-115)

CHANGE SUMMARY: This revision adds "equipotential bonding" to the title and text of the section to clarify that the purpose of the bonding required by this section is "equipotential bonding" rather than the "bonding (bonded)" defined in Article 100. Also, alternate means of bonding have been noted for when other, more traditional means are not available.

2005 CODE: "**680.26 Equipotential Bonding.**

(A) **Performance.** The equipotential bonding required by this section shall be installed to eliminate voltage gradients in the pool area as prescribed.

FPN: ~~This section does not require that~~ The 8 AWG or larger solid copper bonding conductor shall not be required to be extended or attached to any remote panelboard, service equipment, or any electrode.

(B) **Bonded Parts.** The parts specified in 680.26(B)(1) through (B)(5) shall be bonded together.

(1) **Metallic Structural Components.** All metallic parts of the pool structure, including the reinforcing metal of the pool shell, coping stones, and deck, shall be bonded. The usual steel tie wires shall be considered suitable for bonding the reinforcing steel together, and welding or special clamping shall not be required. These tie wires shall be made tight. If reinforcing steel is effectively insulated by an encapsulating nonconductive compound at the time of manufacture and installation, it shall not be required to be bonded. Where reinforcing steel of the pool shell or the reinforcing steel of coping stones and deck is encapsulated with a nonconductive compound or another conductive material is not available, provisions shall be made for an alternate

FIGURE 6.15

means to eliminate voltage gradients that would otherwise be provided by unencapsulated, bonded reinforcing steel.

(2) Underwater Lighting. All metal forming shells and mounting brackets of no-niche luminaires (fixtures) shall be bonded unless a listed low-voltage lighting system with nonmetallic forming shells not requiring bonding is used.

(3) Metal Fittings. All metal fittings within or attached to the pool structure shall be bonded. Isolated parts that are not over 100 mm (4 in.) in any dimension and do not penetrate into the pool structure more than 25 mm (1 in.) shall not require bonding.

(4) Electrical Equipment. Metal parts of electrical equipment associated with the pool water circulating system, including pump motors and metal parts of equipment associated with pool covers, including electric motors, shall be bonded. Accessible metal parts of listed equipment incorporating an approved system of double insulation and providing a means for grounding internal non-accessible, non-current-carrying metal parts shall not be bonded by a direct connection to the equipotential bonding grid. The means for grounding internal non-accessible, non-current-carrying metal parts shall be an equipment grounding conductor run with the power-supply conductors in the case of motors supplied with a flexible cord, or a grounding terminal in the case of motors intended for permanent connection.

Where a double-insulated water-pump motor is installed under the provisions of this rule, a solid 8 AWG copper conductor that is of sufficient length to make a bonding connection to a replacement motor shall be extended from the bonding grid to an accessible point in the motor vicinity. Where there is no connection between the swimming pool bonding grid and the equipment grounding system for the premises, this bonding conductor shall be connected to the equipment grounding conductor of the motor circuit.

(5) Metal Wiring Methods and Equipment. Metal-sheathed cables and raceways, metal piping, and all fixed metal parts that are within the following distances of the pool, except those separated from the pool by a permanent barrier, shall be bonded ~~that are within the following distances of the pool:~~

(1) Within 1.5 m (5 ft) horizontally of the inside walls of the pool

(2) Within 3.7 m (12 ft) measured vertically above the maximum water level of the pool, or any observation stands, towers, or platforms, or any diving structures

(C) Equipotential ~~Common~~ Bonding Grid. The parts specified in 680.26(B) shall be connected to an equipotential ~~a common~~ bonding grid with a solid copper conductor, insulated, covered, or bare, not smaller than 8 AWG or rigid metal conduit of brass or other identified corrosion-resistant metal conduit. Connection shall be made by exothermic welding or by listed pressure connectors or clamps that are labeled as being suitable for the purpose and are of stainless steel, brass, copper, or copper alloy. The equipotential common bonding grid shall extend under paved walking surfaces for 1 m (3 ft.) horizontally

<u>beyond the inside walls of the pool and shall</u> be permitted to be any of the following:

(1) Structural Reinforcing. The structural reinforcing steel of a concrete pool where the reinforcing rods are bonded together by the usual steel tie wires or the equivalent

(2) Bolted or Welded. The wall of a bolted or welded metal pool

(3) <u>Alternate means. This system shall be permitted to be constructed as specified in (a) through (c):</u>

    a. <u>Materials and Connections. The grid shall be constructed of minimum 8 AWG bare</u> ~~A~~ solid copper conductor <u>conductors.</u> ~~insulated, covered, or bare, not smaller than 8 AWG.~~ <u>Conductors shall be bonded to each other at all points of crossing. Connection should be made as required by 680.26(D).</u>

    b. <u>Grid Structure. The equipotential bonding grid shall cover the contour of the pool and the pool deck extending 1 m (3 ft.) horizontally from the inside walls of the pool. The equipotential bonding grid shall be arranged in a 300 mm (12 in.) by 300 mm (12 in.) network of conductors in a uniformly spaced perpendicular grid pattern with tolerance of 100 mm (4 in.).</u>

    c. <u>Securing. The below-grade grid shall be secured within or under the pool and deck media.</u>

~~(4) Rigid metal conduit or intermediate metal conduit of brass or other identified corrosion-resistant metal conduit~~

(D) Connections. Where structural reinforcing steel or the walls of bolted or welded metal pool structures are used as <u>an equipotential</u> ~~a common~~ bonding grid for non-electrical parts, the connections shall be made in accordance with 250.8."

CHANGE SIGNIFICANCE: As can be seen, several significant changes have been made to this section. The first and most obvious change is that the title of the section has been changed to "Equipotential Bonding." This change clarifies the focus of this section, which is bonding for the purpose of equalizing potentials (voltage gradients) in and around the pool area.

The Fine Print Note following 680.26(A) has been changed to a requirement to indicate that the bonding conductor is not required to be extended to or attached to remote panel boards, service equipment, or any grounding electrode. This statement emphasizes that the purpose of the bonding conductor is to equalize potentials and voltage gradients in the pool area and is not installed for the purpose of carrying current in the case of line-to-ground faults. The equipment grounding conductor is installed for that purpose.

In Subsection (C), a rigid metal conduit of brass or other identified corrosion-resistant metal conduit is recognized as a connection means for the equipotential bonding grid. Also, the equipotential bonding grade is required to extend under paved walking services for not less than 3 feet horizontally beyond the inside walls of the pool.

An alternate means for creating the equipotential bonding grid is provided in (C)(3). It should be used where reinforcing steel is encapsulated or is not otherwise conductive and thus could not be used as the equipotential bonding grid. The alternate means consists of 8 AWG solid copper conductors installed in a grid format 12 inches by 12 inches with a tolerance of 4 inches. This means that the grid could be from 8 inches by 8 inches to as much as 16 inches by 16 inches. The copper conductors are required to be bonded to each other at all points of crossing in a manner provided for in 680.26(D).

Section 250.62(D) refers to connections made in accordance with 250.8. Section 250.8 requires that connections be made by exothermic welding, listed pressure connectors, listed clamps, or other listed means. The listing requirements will ensure that connectors are suitable for concrete encasement. Listed connectors may be identified for concrete encasement by being marked "suitable for direct burial," "direct burial," or "db."

This alternate equipotential bonding grade is required to extend not less than 3 feet horizontally from the inside walls of the pool, presumably to bond the walking areas around the pool.

680.32, 680.34

Storable Pools, Ground-Fault Circuit Interrupters Required, Receptacle Locations

NEC page 538

FIGURE 6.16

CHANGE TYPE: Revision and New (Proposal 17-132a)

CHANGE SUMMARY: This proposal returns the requirements for ground-fault circuit interrupter (GFCI) protection of receptacles to the Code for storable pools.

2005 CODE: "680.32 Ground-Fault Circuit Interrupters Required. All electrical equipment, including power-supply cords, used with storable pools shall be protected by ground-fault circuit interrupters.

All 125-volt receptacles located within 6.0 m (20 ft.) of the inside walls of a storable pool shall be protected by a ground-fault circuit interrupter. In determining these dimensions, the distance to be measured shall be the shortest path the supply cord of an appliance connected to the receptacle would follow without piercing a floor, wall, ceiling, doorway with hinged or sliding door, window opening, or other effective permanent barrier.

FPN: For flexible cord usage, see 400.4."

"680.34 Receptacle Locations. Receptacles shall not be less than 3.0 m (10 ft) from the inside walls of a pool. In determining these dimensions, the distance to be measured shall be the shortest path the supply cord of an appliance connected to the receptacle would follow without piercing a floor, wall, ceiling, doorway with hinged or sliding door, window opening, or other effective permanent barrier."

CHANGE SIGNIFICANCE: The requirements for ground-fault circuit interrupter protection of receptacles that supply power to a storable pool were inadvertently left out during the rewrite of Article 680 for the 2002 NEC. Section 680.34, which gives the requirement for receptacles in the vicinity of the storable pool, is new to this part of Article 680.

Note that both Section 680.32 and Section 680.34 do not require the installation of a dedicated branch circuit for the storable pool, but instead regulate the distance receptacles are to be located from the pool as well as require the receptacles within 20 feet of the pool to have GFCI protection. Since storable pools are most commonly placed on the property after the structure is occupied, the spacing rules and GFCI requirements are controlled by the location of the storable pool to existing receptacles.

CHANGE TYPE: Revision (Proposal 17-143)

CHANGE SUMMARY: This revision expands the requirements for GFCI protection for equipment related to fountains.

2005 CODE: "(A) Ground-Fault Circuit Interrupter. <u>Luminaires (lighting fixtures), submersible pumps, and other submersible equipment,</u> ~~Fountain equipment,~~ unless listed for operation at 15 volts or less and supplied by a transformer that complies with 680.23(A)(2), shall be protected by a ground-fault circuit interrupter."

CHANGE SIGNIFICANCE: The previous "fountain equipment" has been replaced with "luminaires (lighting fixtures), submersible pumps, and other submersible equipment" to more explicitly explain the kind of equipment required to have GFCI protection unless listed for operation at fifteen volts or less and supplied by a transformer that complies with the rules in 680.23(A)(2). In that section, relevant transformers are required to be listed as swimming pool and spa transformers, and must have an isolated winding with an ungrounded secondary and a grounded metal barrier between the primary and secondary windings.

680.51(A)
Fountains, Ground-Fault Circuit Interrupter
NEC page 540

FIGURE 6.17

680.74

Hydromassage Bathtubs, Bonding

NEC page 543

CHANGE TYPE: Revision (Proposal 17-153; Comment 17-183)

CHANGE SUMMARY: Revisions are made regarding the bonding rules for hydromassage bathtubs. The previous rule on double-insulated pumps has been deleted.

2005 CODE: "**680.74 Bonding.** All metal piping systems <u>and all grounded</u> metal parts <u>in contact with the circulating water</u> ~~of electrical equipment, and pump motors associated with the hydromassage tub~~ shall be bonded together using a copper bonding jumper, insulated, covered, or bare, not smaller than 8 AWG solid. ~~Metal parts of listed equipment incorporating an approved system of double insulation and providing a means for grounding internal non-accessible, non- current-carrying metal parts shall not be bonded.~~"

CHANGE SIGNIFICANCE: The issue of whether the double-insulated pumps used for hydromassage bathtub should be bonded received a lot of attention at the proposal and comment stage of the recent Code development process. The sentence to prohibit bonding double-insulated motors has been deleted, partly due to the requirements on double-insulated bathtub pumps and their mounting in UL 11795.

With the change to the first sentence, only equipment that is grounded and in contact with the circulating water of the hydromassage bathtub is required to be grounded. If the pump is double-insulated, and supplied with a two-wire cord, it will not be grounded and thus is not required to be bonded.

Hydromassage bathtub motors are not accessible by bathtub occupants and are required by UL 1795 to have their live parts above the mounting surface in the event of a leak. The motors also must isolate

FIGURE 6.18

any internal metal parts that might become energized in a failure (the motor shaft in particular) from the water. However, an internal failure of the motor would not produce the same hazards as an outdoor storable pool unit that is accessible and may have wet surfaces. The grounding of internal dead metal parts, therefore, is not needed.

Article 682

Natural and Artificially Made Bodies of Water

NEC page 543

FIGURE 6.19

CHANGE TYPE: New (Proposal 17-154; Comment 17-184)

CHANGE SUMMARY: A new Article has been added covering natural and artificially made bodies of water.

2005 CODE: "ARTICLE **682 Natural and Artificially Made Bodies of Water**

I. General

682.1 Scope. This article applies to the installation of electrical wiring for and equipment in and adjacent to natural or artificially made bodies of water not covered by other articles in this *Code*, such as but not limited to aeration ponds, fish farm ponds, storm retention basins, treatment ponds, irrigation (channels) facilities.

CHANGE SIGNIFICANCE: This proposal originated with the NEC CMP-17 Task Group on Other Bodies of Water, which was formed by an action of the 2002 NEC CMP-20's direction to investigate the need for a new article to cover the types of installations that include bodies of water involving electrical equipment not covered by Article 680 in the NEC. (This direction is referenced in NEC May 2001 ROC Comments 20-12 and 20-13 and NEC May 2001 ROP Proposals 20-32 and 20-34).

This change also takes into account the statement of 1996 NEC CMP-20 upon their rejection of NFPA 70 A95 ROP Proposal 20-53 for 1993 NEC Section 680-4 to include ponds, which indicated in part: "Storm retention basins, sewage treatment ponds, and similar bodies of water are not covered under the scope of Article 680. It would be impractical to require these facilities to comply with Part E of Article 680. The term 'pond' is not even referred to in Part E covering fountains."

Several of the definitions provided are unique to this article. "Natural bodies of water" include lakes, streams, ponds, rivers, and other naturally occurring bodies of water which may vary in depth throughout the year. "Artificially Made Bodies of Water" include bodies of water that have been constructed or modified to fit some decorative or commercial purpose such as, but not limited to aeration ponds, fish farm ponds, storm retention basins, treatment ponds, irrigation (channels) facilities. Water depths may vary seasonally or be controlled.

682.3 Other Articles. Wiring and equipment is required to comply with other articles of the Code except as modified by this article. If the water is subject to boat traffic, the wiring must comply with 555.13(B).

682.10 Electrical Equipment and Transformers. No portion of an enclosure for electrical equipment not identified for operation while submerged is permitted to be located below the electrical datum plane.

682.11 Location of Service Equipment. The service equipment for floating structures and submersible electrical equipment are required to be located no closer than 5 ft. horizontally from the shoreline. Live parts must be elevated a minimum of 12 in. above the electrical datum plane. *Live parts* is defined in Article 100 as "energized conductive components."

It is required that service equipment disconnect (itself) when the water level reaches the height of the established electrical datum plane. This will require some type of water level sensor and a shunt-trip circuit breaker, or solenoid-operated switch or a switch with a relay.

682.12 Electrical Connections. All electrical connections not intended for operation while submerged are required to be located at least 12 in. above the deck of a floating or fixed structure, but not below the electrical datum plane.

682.13 Wiring Methods and Installation. The wiring methods and installations of Chapter 3 and Articles 553, 555, and 590 are permitted to be used if they are identified for use in wet locations.

682.14 Disconnecting Means for Floating Structures or Submersible Electrical Equipment.

(A) Type. The disconnecting means is permitted to consist of a circuit breaker, switch, or both and is required to be properly identified as to which structure or equipment it controls.

(B) Location. The disconnecting means must be readily accessible on land and located in the supply circuit ahead of the structure or the equipment connection. The disconnecting means is required to be located not more than 30 in. from the structure or equipment connection. The disconnecting means also is required to be within sight of but not closer than 5 ft. horizontally from the edge of the shoreline and live parts and elevated a minimum of 12 in. above the electrical datum plane.

682.15 Ground Fault Circuit Interrupter (GFCI) Protection. GFCI protection is required for:

- 15- and 20-ampere single-phase 125-volt through 250-volt receptacles

- For those installed outdoors (the section does not indicate how far away from the shoreline this rule applies to)

- In or on floating buildings or structures within the electrical datum plane area that are used for storage, maintenance, or repair where portable electric hand tools, electrical diagnostic equipment, or portable lighting equipment are to be used.

The GFCI protection device shall be located not less than 12 in. above the established electrical datum plane. See 682.33(B) where all circuits 60-amperes or less operating at 120 through 250 volts are required to have GFCI protection.

682.30 Grounding. Wiring and equipment within the scope of this article are required to be grounded as specified in Articles 250, 553, and 555 and with the requirements in this Part III (Sections 682.31 through 33).

682.31 Equipment Grounding Conductors.

(A) Type. Equipment grounding conductors are required to be insulated copper conductors sized in accordance with 250.122 but not smaller than 12 AWG.

(B) Feeders. Where a feeder supplies a remote panel board, an insulated equipment grounded conductor is required to extend from a grounding terminal in the service to a grounding terminal and busbar in the remote panel board.

(C) Branch Circuits. The insulated equipment grounding conductor for branch circuits is required to terminate at a grounding terminal in a remote panel board or the grounding terminal in the main service equipment.

(D) Cord- and Plug-Connected Appliances. Where required to be grounded, cord- and plug-connected appliances shall be grounded by means of an equipment grounding conductor in the cord and a grounding-type attachment plug.

682.32 Bonding of Non-Current-Carrying Metal Parts. The following parts are required to be bonded to the grounding bus in the panelboard:

- All metal parts in contact with the water
- All metal piping, tanks
- All non-current-carrying metal parts that may become energized

682.33 Equipotential Planes and Bonding of Equipment Planes. An equipotential plane is required to be installed where required in this section to mitigate step and touch voltages at electrical equipment. The term *equipotential plane* as used in this section is defined in 682.2 as "Equipotential Plane. An area where wire mesh or other conductive elements are on, embedded in, or placed under the walk surface within 75mm (3 in.), bonded to all metal structures and fixed nonelectrical equipment that may become energized, and connected to the electrical grounding system to prevent a difference in voltage from developing within the plane."

(A) Equipotential plane required:

- Installed adjacent to all outdoor service equipment or disconnecting means that control equipment in or on the water that have a metallic enclosure and controls accessible to personnel and likely to become energized
- The equipotential plane must encompass the area around the equipment and shall extend from the area directly below the equipment out not less than 900 mm (36 in.) in all directions from which a person would be able to stand and come in contact with the equipment.

(B) Equipotential plane not required for:

- The controlled equipment supplied by the service equipment or disconnecting means
- All circuits rated not more than 60 amperes at 120 through 250 volts, single phase, are required to have GFCI protection.

(C) Equipotential planes are required to be:

- Bonded to the electrical grounding system with a solid copper conductor, insulated, covered or bare, and not smaller than 8 AWG
- Connections are required to be made by exothermic welding or by listed pressure connectors or clamps that are labeled as being suitable for the purpose and are of stainless steel, brass, copper, or copper alloy.

685.1

Integrated Electrical Systems, Scope

NEC page 544

CHANGE TYPE: Revision (Proposal 12-89; Comment 12-49)

CHANGE SUMMARY: For integrated electrical systems, conditions have been added for determining that qualified persons are actually performing the work.

2005 CODE: "**685.1 Scope.** This article covers integrated electrical systems, other than unit equipment, in which orderly shutdown is necessary to ensure safe operation. An *integrated electrical system* as used in this article is a unitized segment of an industrial wiring system where all of the following conditions are met:

(1) An orderly shutdown is required to minimize personnel hazard and equipment damage.

(2) The conditions of maintenance and supervision ensure that qualified persons service the system. The name(s) of the qualified person(s) shall be kept in a permanent record at the office of the establishment in charge of the completed installation. A person designated as a qualified person shall possess the skills and knowledge related to the construction and operation of electrical equipment and installation and shall have received documented safety training on the hazards involved. Documentation of their qualifications shall be on file with the office of the establishment in charge of the completed installation.

(3) Effective safeguards, acceptable to the authority having jurisdiction, are established and maintained."

FIGURE 6.20

CHANGE SIGNIFICANCE: This proposal maintained that the NEC is primarily a prescriptive code, yet "where the conditions of maintenance and supervision ensure that only qualified persons service the installation" is only a performance requirement. Without requirements indicating (a) whether the qualified person is an employee of the owner of the premises or is a separately contracted person and (b) whether the qualified person has been verified by the authority having jurisdiction as meeting the definition of a qualified person as shown in the definitions of this Code, no prescriptive requirements have been mandated. Changes made to this section are intended to qualify "The conditions of supervision and maintenance ensure that qualified persons service the system."

690.2, 690.31(E)

Building Integrated Photovoltaics, Wiring Methods Permitted

NEC pages 546, 551

CHANGE TYPE: New Definition and Section (Proposal 13-23)

CHANGE SUMMARY: A new definition for *building integrated photovoltaics* has been added to 690.2, and the term is used in new Section 690.31(E). Building-integrated photovoltaics are installed as a part of the building or structure outer surface.

2005 CODE: *"690.2 Building-Integrated Photovoltaics. Photovoltaic cells, devices, modules, or modular materials that are integrated into the outer surface or structure of a building and serve as the outer protective surface of the building."*

"**690.31(E). Direct-Current Photovoltaic Source and Output Circuits Inside a Building.** Where direct current photovoltaic source or output circuits of a utility-interactive inverter from a building-integrated or other photovoltaic system are run inside a building or structure, they shall be contained in metallic raceways or enclosures from the point of penetration of the surface of the building or structure to the first readily accessible disconnecting means. The disconnecting means shall comply with 690.14(A) through 690.14(D)."

CHANGE SIGNIFICANCE: Building-integrated photovoltaic systems are defined in a new definition in 690.2. These are combination photovoltaic systems and also serve as a building or structure outer finish material, often taking the form of siding or roofing materials. Listed products are available, and more are expected to be commercially available in the future.

Integrated photovoltaic systems may have multiple penetration points in the surface of a building or structure. In even a small residential system, the penetrations may number in the hundreds. It is therefore not possible to install an accessible disconnect at each point of penetration. Wiring these dc PV source and output circuits or other dc source and output circuits from roof-mounted PV arrays in metallic raceways or metallic enclosures from the point of penetration to the

FIGURE 6.21

first readily accessible disconnect will meet the intent of the NEC for electrical shock and fire hazard safety for these circuits. Metallic raceways and enclosures provide greater mechanical protection and fire hazard reduction than do non-metallic raceways; enclosures for these circuits may be energized any time the PV materials are exposed to light.

690.14(D)

Utility-Interactive Inverters Mounted in Not-Readily Accessible Locations

NEC page 550

CHANGE TYPE: New (Proposal 13-36; Comments 13-36 and 13-37)

CHANGE SUMMARY: A new section has been added for utility-interactive inverters, addressing requirements for the location, disconnecting means, conductors, and corresponding plaque.

2005 CODE: "**690.14(D) Utility-Interactive Inverters Mounted in Not-Readily Accessible Locations.** Utility-interactive inverters shall be permitted to be mounted on roofs or other exterior areas that are not readily accessible. These installations shall comply with (1) through (4).

(1) A direct-current photovoltaic disconnecting means shall be mounted within sight of or in the inverter.

(2) An alternating-current disconnecting means shall be mounted within sight of or in the inverter.

(3) The alternating-current output conductors from the inverter and an additional alternating-current disconnecting means for the inverter shall comply with 690.14(C)(1).

(4) A plaque shall be installed in accordance with 705.10."

CHANGE SIGNIFICANCE: An *interactive system* is defined in 690.2 as "A solar photovoltaic system that operates in parallel with and may deliver power to an electrical production and distribution network. For the purpose of this definition, an energy storage subsystem of a solar photovoltaic system, such as a battery, is not another electrical production source."

Several utility-interactive inverters have outdoor-rated enclosures. These inverters may be mounted near the photovoltaic array and may

FIGURE 6.22

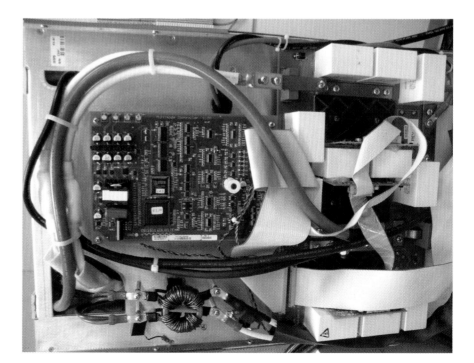

even be mounted on the roofs of dwellings or other buildings in areas that are not readily accessible. AC and dc disconnecting means are still required to service the inverter and should be located near or in the inverter, but they do not need to be readily accessible. The ac outputs of these inverters should be considered as outputs of the photovoltaic system, and these circuits should be treated the same as any other supply for the building; e.g., the routing of the conductors and the location of the disconnect should be established by 690.14(C)(1).

This type of installation may even be safer than an installation where the dc photovoltaic source circuits have to be run some distance from photovoltaic array to a remotely located, readily accessible dc disconnect located near the inverter. In the case of an inverter mounted near the photovoltaic array, opening the readily accessible ac disconnect ensures that all conductors except the dc photovoltaic source and output conductors (near the photovoltaic array) are unenergized.

690.47(C)

Photovoltaic Systems With Alternating-Current and Direct-Current Grounding Requirements

NEC page 553

CHANGE TYPE: New (Proposal 13-47)

CHANGE SUMMARY: A new section has been added to Article 690 to provide rules for sizing grounding electrode conductors and installing grounding electrodes for ac and dc systems from solar photovoltaic power systems.

2005 CODE: "690.47(C) Systems with Alternating-Current and Direct-Current Grounding Requirements. Photovoltaic power systems with both alternating-current (ac) and direct-current (dc) grounding requirements shall be permitted to be grounded as described in (1) or (2).

(1) A grounding-electrode conductor shall be connected between the identified dc grounding point to a separate dc grounding electrode. The dc grounding electrode conductor shall be sized according to 250.166. The dc grounding electrode shall be bonded to the ac grounding electrode to make a grounding electrode system according to 250.52 and 250.53. The bonding conductor shall be no smaller than the largest grounding electrode conductor, either ac or dc.

(2) The dc grounding electrode conductor and ac grounding electrode conductor shall be connected to a single grounding electrode. The separate grounding electrode conductors shall be sized as required by 250.66(ac) and 250.166(dc)."

CHANGE SIGNIFICANCE: The NEC previously addressed the grounding of ac systems and dc systems in separate sections in Article 250, but it did not specifically address systems where both dc and ac grounding

FIGURE 6.23

must be made from the same equipment. In photovoltaic systems there are usually both dc and ac circuits/systems that need to be grounded.

If the equipment-grounding system is connected in such a manner that no ground-fault currents can normally flow through the grounding electrode conductor, then the intent of the Code is met. Grounding electrode conductors and grounding electrodes are not intended to carry ground-fault currents, only lightning-induced surge currents and currents from accidental cross connection to other systems. If no ground-fault currents travel in grounding electrode conductors, then the sizing requirements apply from NEC Sections 250.66 for ac systems and 250.166 for dc systems.

This addition allows two clearly specified approaches to providing ac and dc system grounding. It also eliminates earlier confusion surrounding techniques found in Article 250.

692.1

Fuel Cell Systems, Scope

NEC page 556

CHANGE TYPE: Revision (Proposal 13-67)

CHANGE SUMMARY: A sentence has been added to the scope of Article 692 to indicate that the output of a fuel cell system may be either alternating current (ac) or direct current (dc).

2005 CODE: "**692.1 Scope.** This article identifies the requirements for the installation of fuel cell power systems, which may be stand-alone or interactive with other electrical power production sources and may be with or without electrical energy storage such as batteries. These systems may have ac or dc output for utilization."

CHANGE SIGNIFICANCE: Fuel cells have a dc output, which may be inverted into an ac output or converted (via a dc-dc converter) to a usable dc output.

All fuel systems create direct current. The output of fuel cell systems associated with premises wiring systems is routed through an inverter and power conditioner to produce a manufactured alternating current sine wave to be compatible with 60-Hz electrical systems commonly in use in the United States. The change to the scope of this article (with wording taken from 690.1) recognizes that there are many legitimate uses for direct-current output from fuel cells.

FIGURE 6.24

CHANGE TYPE: Revision (Proposal 13-83; Comment 13-50)

CHANGE SUMMARY: A revision has been made regarding the connection permitted to be made on the supply side of a normal building or structure service disconnecting means.

2005 CODE: "**695.3(A)(1) Electric Utility Service Connection.** A fire pump shall be permitted to be supplied by a separate service, or <u>from a connection</u> ~~by a tap~~ located ahead of and not within the same cabinet, enclosure, or vertical switchboard section as the service disconnecting means. The connection shall be located and arranged so as to minimize the possibility of damage by fire from within the premises and from exposing hazards. A tap ahead of the service disconnecting means shall comply with 230.82<u>(5)</u> ~~(4)~~. The service equipment shall comply with the labeling requirements in 230.2 and the location requirements in 230.72(B). [<u>NFPA 20:9.2.2</u>]"

CHANGE SIGNIFICANCE: The substantiation for this proposal indicated that the connection being made on the supply side of normal service equipment is more properly termed a "connection" rather than a "tap." The Code Panel accepted the proposed change but pointed out that the definition of *tap* in 240.2 only applies to Article 240. The word *tap* continues to be used in the penultimate sentence in this section.

The location of the service disconnecting means for a fire pump is required to comply with 230.72(B), which states, "(B) Additional Service Disconnecting Means. The one or more additional service disconnecting means for fire pumps, for legally required standby, or for optional standby services permitted by 230.2 shall be installed remote from the one to six service disconnecting means for normal service to minimize the possibility of simultaneous interruption of supply." While the Code uses the phrase, "installed remote from the one to six service disconnecting means," no specific separation distance is provided. The authority having jurisdiction should be consulted for any local regulations or interpretations.

695.3(A) (1)

Electric Utility Service Connection for Fire Pumps
NEC page 559

FIGURE 6.25

695.4(B)(1) and (B)(2)(3)

Supervised Connection, Overcurrent Device Selection, Disconnecting Means

NEC page 561

CHANGE TYPE: Revision and New (Proposals 13-85 and 13-86)

CHANGE SUMMARY: 695.4(B)(1) has been changed to emphasize that the requirement to carry the locked-rotor currents indefinitely applies only to the overcurrent devices for the fire pump motor circuit. The change to (B)(2)(3) requires that the disconnecting means for fire pumps are not located within equipment that feeds loads other than the fire pump.

2005 CODE: "695.4(B)(1) Overcurrent Device Selection. The overcurrent protective device(s) shall be selected or set to carry indefinitely the sum of the locked-rotor current of the fire pump motor(s) and the pressure maintenance pump motor(s) and the full-load current of the associated fire pump accessory equipment when connected to this power supply. <u>The requirement to carry the locked-rotor currents indefinitely shall not apply to conductors or devices other than overcurrent devices in the fire pump motor circuit(s).</u>

(2) Disconnecting Means. The disconnecting means shall comply with all of the following:

(1) Be identified as suitable for use as service equipment

(2) Be lockable in the closed position

(3) <u>Not be located within equipment that feeds loads other than the fire pump</u>

(4) Be located sufficiently remote from other building or other fire pump source disconnecting means such that inadvertent contemporaneous operation would be unlikely"

CHANGE SIGNIFICANCE: The proposal to revise 695.4(B)(1) was prompted by the fact that installers were being required to apply the locked-rotor rules for overcurrent protective devices to carry indefinitely the sum of the locked-rotor current of the fire pump motor(s),

FIGURE 6.26

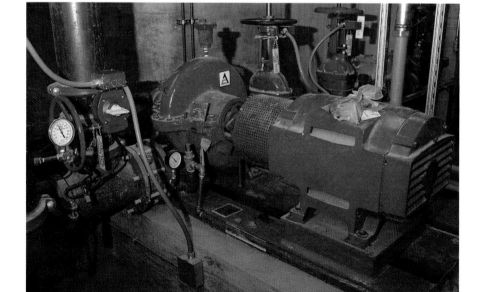

the pressure maintenance pump motor(s), the full-load current of the associated fire pump accessory equipment, and in some cases the remaining loads supplied by the transformer. The locked-rotor rules now apply only to the overcurrent device(s) for the fire pump motor circuit(s).

In 695.4(B)(2), it was previously stated that equipment manufacturers are requested to provide various panelboard and switchboard custom configurations from customers located across the country. Our experience has found that the wording "located sufficiently remote" in the present 695.4(B)(2)(3) has a varying degree of interpretation across the country and has been understood to mean:

(1) located within a panelboard feeding various loads with the fire pump disconnect located at the bottom of the panel (sufficiently remote?)

(2) located within a switchboard structure feeding various loads either at the bottom of a single section (sufficiently remote?) or in a separate section of an assembly (sufficiently remote?)

(3) located as a separate enclosure

Including a new item (3) will clarify that a fire pump disconnect is not permitted in a panelboard or switchboard that feeds other loads in order to ensure continuity of power and further support the appropriate interpretation of "sufficiently remote" in newly numbered item (4).

695.5(B), (C)(2)

Overcurrent Protection for Transformers, Overcurrent Protection for Feeders

NEC pages 560, 561

CHANGE TYPE: (Proposals 13-88 and 13-89; Comments 13-52a and 13-52b)

CHANGE SUMMARY:

2005 CODE: "**695.5(B) Overcurrent Protection.** The primary overcurrent protective device(s) shall be selected or set to carry indefinitely the sum of the locked-rotor current of the fire pump motor(s) and the pressure maintenance pump motor(s) and the full-load current of the associated fire pump accessory equipment when connected to this power supply. Secondary overcurrent protection shall not be permitted. The requirement to carry the locked-rotor currents indefinitely shall not apply to conductors or devices other than the overcurrent devices in the fire pump motor circuit(s)."

"**695.5(C)(2) Overcurrent Protection.** The transformer size, the feeder size, and the overcurrent protective device(s) shall be coordinated such that overcurrent protection is provided for the transformer in accordance with 450.3 and for the feeder in accordance with 215.3, and such that the overcurrent protective device(s) is selected or set to carry indefinitely the sum of the locked-rotor current of the fire pump motor(s), the pressure maintenance pump motor(s), the full-load current of the associated fire pump accessory equipment, and 100 percent of the remaining loads supplied by the transformer. The requirement to carry the locked-rotor currents indefinitely shall not apply to conductors or devices other than the overcurrent devices in the fire pump motor circuit(s)."

CHANGE SIGNIFICANCE: Section 695.5 applies to transformers used for fire pumps; 695.5(B) contains overcurrent protection rules for these transformers. The overcurrent device on the primary is required to carry

FIGURE 6.27

indefinitely the sum of the locked-rotor current of the fire pump motor(s) and the pressure maintenance pump motor(s) and the full load current of the associated fire pump accessory equipment when connected to this power supply. The locked-rotor current of typical squirrel-cage motors is usually assumed to be about six times the full-load current, though this value can vary depending upon the design of the motor.

The requirement to size overcurrent protective devices to carry indefinitely the sum of the locked-rotor current of the fire pump motor(s) and the pressure maintenance pump motor(s), the full-load current of the associated fire pump accessory equipment, and in some cases the remaining loads supplied by the transformer was being applied incorrectly to conductors and other devices in the fire pump motor circuits. This revision is intended to clarify the locked-rotor-current requirements apply only to the overcurrent protection and not to other conductors or devices in the circuit.

695.6(C) (2), (D), (E), and (H)

Power Wiring for Fire Pumps, Overcurrent Protection, Pump Wiring, and Ground-Fault Protection Prohibition

NEC page 561, 562

CHANGE TYPE: Revision and New (Proposal 13-100)

CHANGE SUMMARY: Conductors supplying only a fire pump motor are now required to comply with 430.22. Larger conductors may be necessary to satisfy the requirement in 695.7. The rules on overload protection for tap conductors for fire pumps have also been revised. Additionally, the wiring methods for fire pumps now include jacketed Type MC cable, and a new section prohibits ground-fault protection of equipment for fire pump supply conductors.

2005 CODE: "**695.6(C)(2) Fire Pump Motors Only.** Conductors supplying only a fire pump motor shall have a <u>minimum ampacity in accordance with 430.22 and shall comply with the voltage drop requirements in 695.7</u> ~~rating not less than 125 percent of the fire pump motor(s) full-load current(s)~~.

    (D) Overload Protection. Power circuits shall not have automatic protection against overloads. ~~Except as provided in 695.5(C)(2),~~ Branch-circuit and feeder conductors shall be protected against short circuit only. Where a tap is made to supply a fire pump, ~~and~~ the ~~tap~~ wiring <u>shall be treated as service conductors</u> ~~is run~~ in accordance with 230.6. The applicable distance and size restrictions in 240.21 shall not apply.

    Exception No. 1: Conductors between storage batteries and the engine shall not require overcurrent protection or disconnecting means.

    Exception No. 2: For on-site standby generator(s) <u>rated to</u> ~~that~~ produce continuous currents in excess of 225 percent of the full-load amperes of the fire pump motor, the conductors between the on-site generator(s) and the combination fire pump transfer switch controller or separately mounted transfer switch shall be installed in accordance with 695.6(B) or protected in accordance with 430.52."

FIGURE 6.28

The protection provided shall be in accordance with the short-circuit current rating of the combination fire pump transfer switch controller or separately mounted transfer switch.

"**(E) Pump Wiring.** All wiring from the controllers to the pump motors shall be in rigid metal conduit, intermediate metal conduit, liquidtight flexible metal conduit, or liquidtight flexible nonmetallic conduit Type LFNC-B, <u>listed Type MC cable with an impervious covering,</u> or Type MI cable."

"**<u>(H) Ground Fault Protection of Equipment.</u>** <u>Ground-fault protection of equipment shall not be permitted for fire pumps.</u>"

CHANGE SIGNIFICANCE: The change to 695.6(C)(2) directs the Code user to 430.22 for the sizing of branch-circuit conductors to the fire pump. That section requires branch-circuit conductors to be sized not less that 125 percent of the full-load current as indicated in the appropriate Tables 430.248 through 430.250. Also, the change directs the user to the existing rule in 695.7, where the requirements for sizing conductors for the maximum voltage drop read: "The voltage at the controller line terminals shall not drop more than 15 percent below normal (controller-rated voltage) under motor starting conditions. The voltage at the motor terminals shall not drop more than 5 percent below the voltage rating of the motor when the motor is operating at 115 percent of the full-load current rating of the motor."

695.6(D) provides rules on overload protection of power circuits. It generally requires that branch circuits and feeder conductors be provided with short-circuit protection only. Changes emphasize that tap conductors are to be treated as service conductors. Service conductors are permitted to be protected against overload only (not overcurrent) by the overcurrent device they terminate in.

695.6(E) provides for the wiring methods permitted for fire pump wiring. Listed Type MC cable with an impervious covering is now an acceptable wiring method. This kind of cable has a continuous PVC outer jacket and is often referred to as "parking deck cable."

Finally, 695.6(H) is a new section and prohibits ground-fault protection of equipment for fire pumps. This is not a new concept—an identical prohibition has been in 215.10, 240.13, and 230.95 for some time.

CHAPTER 7

Special Conditions, Articles 700–780

700.5(B)

Emergency Systems, Selective Load Pickup, Load Shedding, and Peak Load Shaving

NEC page 563

CHANGE TYPE: Revision (Proposal 13-109)

CHANGE SUMMARY: Changes have been made to this section to clarify that the alternate power source is permitted to be used for emergency, legally required standby, and optional standby loads where it has adequate capacity.

2005 CODE: "**(B) Selective Load Pickup, Load Shedding, and Peak Load Shaving.** The alternate power source shall be permitted to supply emergency, legally required standby, and optional standby system loads where the source has adequate capacity or where automatic selective load pickup and load shedding is provided as needed to ensure adequate power to (1) the emergency circuits, (2) the legally required standby circuits, and (3) the optional standby circuits, in that order of priority. The alternate power source shall be permitted to be used for peak load shaving, provided these the above conditions are met.

Peak load-shaving operation shall be permitted for satisfying the test requirement of 700.4(B), provided all other conditions of 700.4 are met.

A portable or temporary alternate source shall be available whenever the emergency generator is out of service for major maintenance or repair."

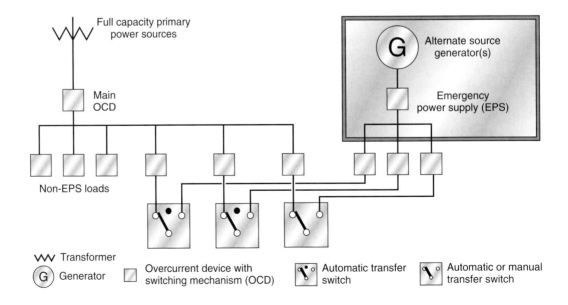

FIGURE 7.1

CHANGE SIGNIFICANCE: Changes made to this section indicate that the alternate power system is permitted to serve the three listed systems if the alternate power system has the capacity. In such a case, load shedding is not required.

It should be noted that 700.6(D) requires that transfer equipment for emergency systems supply only emergency loads. As a result, no fewer than two transfer switches would be required when the emergency system, legally required standby system, and optional standby system are supplied by the same alternate power system such as a generator. One transfer switch would be used for the emergency system and another transfer switch could supply both the legally required standby system and the optional standby system.

700.12 (B)(6)

Emergency System, Outdoor Generator Sets

NEC page 566

CHANGE TYPE: Revision (Proposal 13-125)

CHANGE SUMMARY: This revision changes when a disconnecting means on or at a generator set located within sight of a building or structure is permitted to also serve as the disconnecting means for the conductors that serve or pass through the building or structure.

2005 CODE: "**700.12(B)(6) Outdoor Generator Sets.** Where an outdoor housed generator set is equipped with a readily accessible disconnecting means located within sight of the building or structure supplied, an additional disconnecting means shall not be required where ungrounded conductors <u>serve or</u> pass through the building or structure."

CHANGE SIGNIFICANCE: The new text would permit the disconnecting means for the generator to act as the required disconnecting means for the circuit supplying or passing through a separate building or structure in accordance with 225.31, reportedly the Code Panel's intent in the 2002 revision to this section. The provisions in 225.31 prevent the requirement for an additional disconnecting means to be installed "inside or outside the building or structure ... nearest the point of entrance of the conductors." The conductors noted in 225.31 usually connect to a transfer switch or to emergency system distribution equipment.

FIGURE 7.2

700.12(D)

Emergency Systems, Separate Service
NEC page 566

CHANGE TYPE: Revision (Proposal 13-127)

CHANGE SUMMARY: Changes have been made to this section to emphasize the importance of locating additional services "sufficiently remote" from other services to "minimize the possibility of simultaneous interruption of supply."

2005 CODE: "**700.12(D) Separate Service.** Where acceptable to the authority having jurisdiction as suitable for use as an emergency source of power, an additional a second service shall be permitted. This service shall be in accordance with the applicable provisions of Article 230 and the following additional requirements: with
(1) Separate service drop or lateral
(2) Service conductors sufficiently remote widely separated electrically and physically from any other the normal service conductors to minimize the possibility of simultaneous interruption of supply."

CHANGE SIGNIFICANCE: This revision makes both editorial and substantive changes to this section. The previous language permitted a second service, which was technically incorrect as the emergency service could be the third (or more) service where more than one service is installed as permitted in 230.2.

The requirements for separation of the emergency service from other services have been organized into list form for improved user friendliness as well as for emphasis. The phrase "sufficiently remote" now describes the separation required. While the word *sufficiently* is included in the list of terms in the NEC Style Manual that are vague and

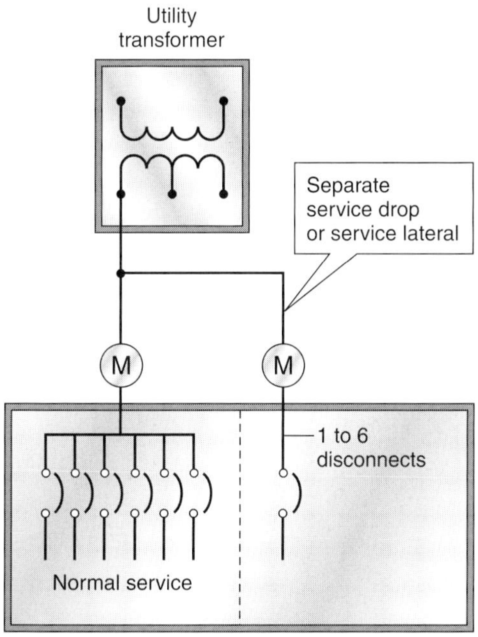

FIGURE 7.3

AHJ may require emergency service to be located
in another room or to have a specific separation

possibly unenforceable, the term is also used in 225.34(B) for required separation of supplies for fire pumps, emergency, legally required standby systems, and optional standby systems. ("Remote from" is used in 230.72(B) for separation of the emergency system.) Due to the use of a possibly unenforceable or vague term, the authority having jurisdiction will have to make a decision about the separation that is acceptable to prevent a failure, such as a fire in other electrical services, from adversely impacting the emergency service.

A similar change is made to 701.11(D) in Proposal 13-142 for legally required standby systems.

CHANGE TYPE: New (Proposal 13-128)

CHANGE SUMMARY: Fuel cell systems have been added to the list of power sources permitted to supply emergency systems. Specific conditions are placed on a fuel cell system in order for it to qualify as a suitable source of power.

2005 CODE: "**700.12(E) Fuel Cell System.** Fuel cell systems used as a source of power for emergency systems shall be of suitable rating and capacity to supply and maintain the total load for not less than two hours of full-demand operation.

Installation of a fuel cell system shall meet the requirements of Parts II through VIII of Article 692.

Where a single fuel cell system serves as the normal supply for the building or group of buildings concerned, it shall not serve as the sole source of power for the emergency standby system."

CHANGE SIGNIFICANCE: Given the present pace of research, development, and testing of fuel cell technology, fuel cells may have matured to the point where such systems may be in general use by the 2005 through 2008 NEC cycle. In the meantime, reliability is matched to already existing sources of power for emergency standby systems.

To be acceptable for use as a source of power for an emergency system, a fuel cell system:

- Must be of suitable rating and capacity to supply and maintain the total load for two hours of full-demand operation.

- Must be installed in compliance with Parts II through VIII of Article 692.

700.12(E)
Emergency Systems, Fuel Cell System
NEC page 566

FIGURE 7.4

- Is not permitted to supply both the normal electrical loads and the emergency system loads from the same unit. (This requirement mandates two fuel cell systems, one for normal power and another for the emergency system.)

Similar text was also accepted for legally required standby systems in 701.11(F) by Proposal 13-143.

CHANGE TYPE: New (Proposals 13-135 and 13-145)

CHANGE SUMMARY: Requirements that emergency system overcurrent devices be selectively coordinated have been added to 700.27. An identical requirement has been added to 701.18 for legally required standby systems.

2005 CODE: "**700.27 Coordination.** Emergency system(s) overcurrent devices shall be selectively coordinated with all supply side overcurrent protective devices."

"**701.18 Coordination.** Legally required standby system(s) overcurrent devices shall be selectively coordinated with all supply side overcurrent protective devices."

CHANGE SIGNIFICANCE: The requirements contained in Article 700 focus on providing a reliable emergency system that will be operational

700.27, 701.18

Emergency Systems and Legally Required Standby Systems, Coordination
NEC pages 567, 570

FIGURE 7.5

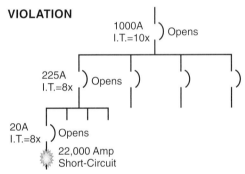

VIOLATION

Fault exceeding the instantaneous trip setting of all 3 circuit breakers in series will open all 3. This will black out the entire system.

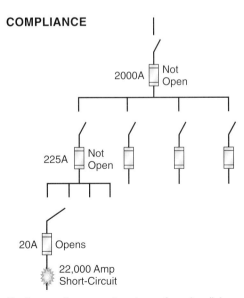

COMPLIANCE

Fault opens the nearest upstream fuse, localizing the fault to the equipment affected. Service to the rest of the system remains energized.

when called upon as supported by the maintenance and testing requirements in 700.4. The critical nature of the dependence on emergency systems for safety of human life, as stated in the scope, supports these requirements.

The concept adopted by accepting this proposal is if overcurrent devices are in series, as is almost always the case with the service, feeder, and branch-circuit overcurrent devices, only the overcurrent device closest to a fault will open. This leaves the remainder of the emergency, or legally required standby system intact and functional.

Article 700 specifically mandates that the emergency circuits be separated from the normal circuits as shown in 700.9(B), and that wiring be specifically located to minimize system hazards as shown in 700.9(C). These requirements reduce the likelihood of faults or failures to the system so it will be operational when called upon. Also, an emphasis is placed on component reliability for emergency lighting for egress. It is imperative that the lighting system remain operational in an emergency. Failure of one component must not result in a condition where a means of egress will be in total darkness (as described in 700.16).

Selectively coordinated overcurrent protective devices will provide a system that will support all of these requirements and principles. With properly selected overcurrent protective devices, a fault in the emergency system will be localized to the overcurrent protective device nearest the fault, allowing the remainder of the system to be functional. This can be accomplished with both fuses and circuit breakers based on the system design and selection of appropriate overcurrent protective devices.

As can be seen above, the same requirement was added in 701.18 (Proposal 13-145) for legally required standby systems.

702.6 Exception

Optional Standby Systems, Transfer Equipment
NEC page 571

CHANGE TYPE: New (Proposal 13-148)

CHANGE SUMMARY: A new exception allows connection of an optional standby alternate power source to equipment in a qualifying industrial facility under specific conditions.

2005 CODE: "**702.6 Transfer Equipment.** Transfer equipment shall be suitable for the intended use and designed and installed so as to prevent the inadvertent interconnection of normal and alternate sources of supply in any operation of the transfer equipment. Transfer equipment and electric power production systems installed to permit operation in parallel with the normal source shall meet the requirements of Article 705.

Transfer equipment, located on the load side of branch circuit protection, shall be permitted to contain supplementary overcurrent protection having an interrupting rating sufficient for the available fault current that the generator can deliver. The supplementary overcurrent protection devices shall be part of a listed transfer equipment.

Transfer equipment shall be required for all standby systems subject to the provisions of this article and for which an electric-utility supply is either the normal or standby source.

Exception: Temporary connection of a portable generator without transfer equipment shall be permitted in industrial installations, with written safety procedures, where conditions of maintenance and supervision insure that only qualified persons service the installation and

FIGURE 7.6

where the normal supply is physically isolated by a lockable disconnect means or by disconnection of the normal supply conductors."

CHANGE SIGNIFICANCE: This new exception permits an optional standby alternate power supply to be connected to an electrical system on a temporary basis without a transfer switch. However, such a connection is limited as follows:

- The connection is "temporary." No time limit is established for the "temporary connection," so the authority having jurisdiction may enforce the time limit in 590.3(A), which limits the temporary wiring to the period of "construction, remodeling, maintenance, repair, or demolition of buildings, structures, equipment, or similar activities."

- The temporary connection is limited to industrial occupancies "with written safety procedures, where conditions of maintenance and supervision insure that only qualified persons service the installation."

- The normal supply is "physically isolated by a lockable disconnect means or by disconnection of the normal supply conductors."

702.7 Exception

Optional Standby Systems, Signals
NEC page 571

CHANGE TYPE: New (Proposal 13-149)

CHANGE SUMMARY: A new exception has been added to indicate that signals for derangement or carrying loads are not required for portable standby sources.

2005 CODE: "**702.7 Signals.** Audible and visual signal devices shall be provided, where practicable, for the following purposes.

(1) **Derangement.** To indicate derangement of the optional standby source.

(2) **Carrying Load.** To indicate that the optional standby source is carrying load.

Exception: Signals shall not be required for portable standby power sources."

CHANGE SIGNIFICANCE: Audible and visual signals are required (where practicable) for emergency systems, legally required standby systems, and permanently installed optional standby systems. The new exception to this section indicates that these signals are not required for portable optional standby power sources, which are usually portable or vehicle mounted generators.

Section 700.7, on emergency systems, and 701.8, on legally required standby systems, have a Fine Print Note that refers to NFPA 110 (*Standard for Emergency and Standby Power Systems*) for information on signals for generator sets. Requirements for indications of derangement include overcrank, high engine temperature, low lube oil pressure, overspeed, air shutdown damper, and remote emergency stop.

FIGURE 7.7

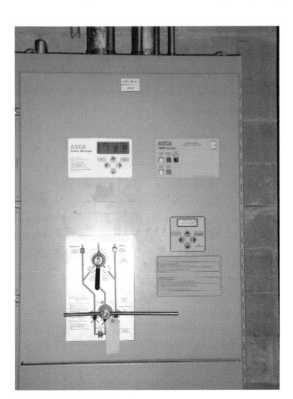

702.11

Optional Standby Systems, Sources of Power

NEC page 571

CHANGE TYPE: New (Proposal 13-152)

CHANGE SUMMARY: This new section permits a disconnecting means on or at an outdoor-housed generator set to serve as the disconnecting means for conductors that enter or serve a building or structure.

2005 CODE: "**702.11 Outdoor Generator Sets.** Where an outdoor housed generator set is equipped with a readily accessible disconnecting means located within sight of the building or structure supplied, an additional disconnecting means shall not be required where ungrounded conductors serve or pass through the building or structure."

CHANGE SIGNIFICANCE: This new Part IV on outdoor generator sets contains provisions identical to Article 700 and 701, permitting the disconnecting means on or at the generator set to serve as the disconnecting means for the conductors that serve or pass through the building or structure.

To be permitted as a disconnecting means for the conductors, the disconnecting means at the generator set must be readily accessible and within sight from the building or structure served. The terms *readily accessible* and *within sight* are defined in Article 100. *Readily accessible* means "capable of being reached quickly for operation, renewal, or inspections without requiring those to whom ready access is requisite to climb over or remove obstacles or to resort to portable ladders, and so forth." The term *within sight* is defined as "Where this Code specifies that one equipment shall be 'in sight from,' 'within sight from,' or 'within sight,' and so forth, of another equipment, the specified equipment is to be visible and not more than 15 m (50 ft) distant from the other."

The new section also permits the disconnecting means for the generator to act as the required disconnecting means for the circuit entering or passing through a separate building or structure as required by 225.31. The same text as provided for generator sets for emergency and legally required standby generators.

FIGURE 7.8

CHANGE TYPE: Revised (Proposal 3-131)

CHANGE SUMMARY: The phrase "shall not be permitted to remain" regarding abandoned Class 1, Class 2, Class 3, and PLTC cable has been replaced with "shall be removed."

2005 CODE: "**725.3(B) Spread of Fire or Products of Combustion.** Section 300.21. The accessible portion of abandoned Class 2, Class 3, and PLTC cables shall <u>be removed</u> ~~not be permitted to remain~~."

CHANGE SIGNIFICANCE: The substantiation for the proposal indicated that the previous language for the removal of abandoned cables was a convoluted way of saying "shall be removed." Though this issue received a lot of attention in the 2005 NEC process, at the Comment stage as well as on the floor of the NFPA Annual Meeting, the language originally accepted by CMP-3 was ultimately accepted.

The definition for *abandoned Class 2, Class 3, and PLTC cable* was added to the 2002 NEC and reads, "Installed Class 2, Class 3, and PLTC cable that is not terminated at equipment and not identified for future use with a tag." The definition was not changed in the 2005 NEC. This remains a contentious issue. All that is required to avoid the time and cost of removing old abandoned cables is to put a tag on the cable identifying it as intended for use in the future.

A change identical to that in 725.3(B) was made to 760.3(A) for fire alarm circuits, to 770.3(A) for optical fiber cables, to 820.3(A) for CATV system cables, and to 830.3(A). The previous language "shall be permitted to remain" was not changed in 800.3(C).

725.3(B)

Spread of Fire or Products of Combustion

NEC page 574

FIGURE 7.9

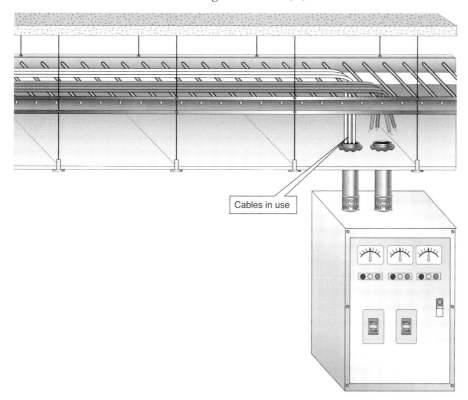

Cables in use

725.3(C)

Ducts, Plenums, and Other Air-Handling Spaces
NEC page 574

CHANGE TYPE: Revision (Proposal 3-133; Comment 3-129)

CHANGE SUMMARY: Editorial changes have been made to the section. Also, plenum signaling raceways can now be installed in "other spaces" used for environmental air.

2005 CODE: "**725.3(C) Ducts, Plenums, and Other Air-Handling Spaces.** ~~Section 300.22 for~~ Class 1, Class 2, and Class 3 circuits installed in ducts, plenums, or other spaces used for environmental air <u>shall comply with 300.22</u>. Type CL2P or CL3P cables <u>and plenum signaling raceways</u> shall be permitted for Class 2 and Class 3 circuits <u>installed in other spaces used for environmental air</u>."

CHANGE SIGNIFICANCE: This is the first of three companion proposals to add non-metallic signaling raceways to Article 725. The same listed raceways are currently found in Articles 770 and 800 for optical Fiber and communication cables. These raceways are commonly being used to facilitate the installation and removal of abandoned cables, since they are suitable for containing signaling cables that are permitted to be installed in spaces used for environmental air.

FIGURE 7.10

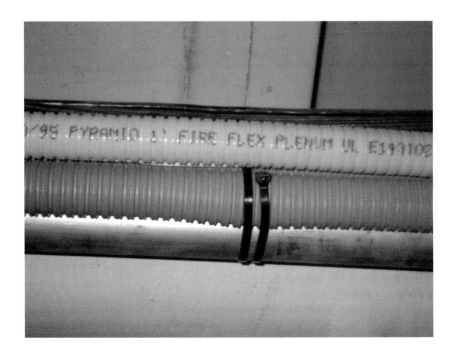

CHANGE TYPE: Revision (Proposal 3-157)

CHANGE SUMMARY: The change to this section clarifies that the marking requirements for Class 2 and Class 3 circuits apply to the equipment supplying the circuits, not to the cables, boxes, and so forth.

2005 CODE: "**725.42 Circuit Marking.** The equipment <u>supplying the circuits</u> shall be durably marked where plainly visible to indicate each circuit that is a Class 2 or Class 3 circuit."

CHANGE SIGNIFICANCE: As previously written, this section was not clear as to which equipment was to be marked to indicate if circuits were Class 2 or Class 3. Since *equipment* has a very broad definition in Article 100, this section has sometimes been interpreted to mean that conduits, boxes, and cables must be marked. Additionally, where substitute cable types are used as permitted by 725.61, this section may even be interpreted to mean that cables must be reidentified as Class 2 or Class 3. Proposal 3-157 intended to make the marking requirements consistent with the reidentification required by 725.52(A), Exception No. 2, and also clarify that requirement.

The Code Panel accepted the clarification; it is now clear that marking required for Class 2 and 3 circuits should be on the equipment at the *source* of the circuit, not necessarily at all other points on the circuit.

725.42

Circuit Marking
NEC page 577

FIGURE 7.11

725.56(F)

Class 2 or Class 3 Conductors or Cables and Audio System Circuits

NEC page 579

CHANGE TYPE: New (Proposal 3-162a)

CHANGE SUMMARY: This new section clearly states that audio system cables are not permitted to be installed in the same cable raceway as Class 2 or Class 3 conductors or cables.

2005 CODE: "**725.56(F) Class 2 or Class 3 Conductors or Cables and Audio System Circuits.** Audio system circuits described in Section 640.9(C) and installed using Class 2 or Class 3 wiring methods in compliance with Sections 725.54 and 725.61 shall not be permitted to be installed in the same cable or raceway with Class 2 or Class 3 conductors or cables."

CHANGE SIGNIFICANCE: The Code Panel developed this proposal that addresses the separation of Class 2 and Class 3 circuits from audio circuits that use Class 2 or Class 3 wiring methods. The audio circuits may have a high enough amperage or voltage that a fault could adversely affect Class 2 or Class 3 circuits, some of which control life safety equipment.

Additionally, Article 725 power sources are restricted to a maximum output of 100 VA in order to be identified as Class 2 or Class 3. Article 725 power sources with outputs greater than 100 VA are required to be installed using Class 1 wiring methods and materials, and Article 725 requires Class 2 and 3 circuits to be separated from Class 1 circuits.

A commercial audio amplifier (Article 640) is not required to use Class 1 wiring methods when the amplifier output is greater than 100 VA. Therefore, it is not possible to know if the audio amplifier's output is equivalent to Class 2 and 3 power source requirements.

FIGURE 7.12

725.61 (D)(2)

Intrinsically Safe Circuits and Nonincendive Field Wiring
NEC page 580

CHANGE TYPE: Revision (Proposal 3-179)

CHANGE SUMMARY: Changes to this section indicate that it is suitable to use circuits derived from Class 2 sources for both intrinsically safe circuits and nonincendive field wiring.

2005 CODE: "**725.61(D)(2) Intrinsically Safe Circuits and Nonincendive Field Wiring.** Wiring for <u>nonincendive circuits as permitted by 501.10(B)(3), and wiring for intrinsically safe circuits as permitted by 504.20,</u> ~~Class 2 circuits as permitted by 501.4(B)(3)~~ shall be permitted <u>for circuits derived from Class 2 sources</u>."

CHANGE SIGNIFICANCE: Circuits derived from Class 2 sources are now permitted as nonincendive circuits as permitted by 501.10(B)(3) and for intrinsically safe circuits as permitted in 504.20.

FIGURE 7.13

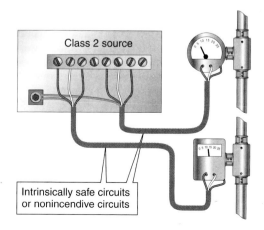

760.21

Non-Power-Limited Fire Alarm Circuit Power Source Requirements

NEC page 585

CHANGE TYPE: Revision (Proposal 3-236)

CHANGE SUMMARY: For non-power-limited fire alarm circuits, the last sentence has been changed to require that "these circuits shall not be supplied through ground-fault circuit interrupters or arc-fault circuit-interrupters."

2005 CODE: "**760.21 NPLFA Circuit Power Source Requirements.** The power source of non-power-limited fire alarm circuits shall comply with Chapters 1 through 4, and the output voltage shall not be more than 600 volts, nominal. These circuits shall not be supplied through ground-fault circuit interrupters <u>or arc-fault circuit interrupters</u>.

FPN: See 210.8(A)(5), Exception No. 3, for receptacles in dwelling-unit unfinished basements that supply power for fire alarm systems."

CHANGE SIGNIFICANCE: As revised for the 2005 NEC, 210.12 continues to require all outlets in dwelling-unit bedrooms to be provided with AFCI protection. This includes 120-volt outlets for smoke detectors. However, the organization of the Code in 90.3 applies to this issue. The rules in Chapter 2, where 210.12 is located, apply generally unless amended or modified by rules in Chapters 5, 6, or 7. Since Article 760 is in Chapter 7 of the Code, the rules in Article 760 can amend the rules in Article 2. As a result, the rule in 760.21 can be interpreted to act as an exception to the rule in 210.12 and prohibit the non-power-limited fire alarm system from being supplied by an AFCI-protected branch circuit.

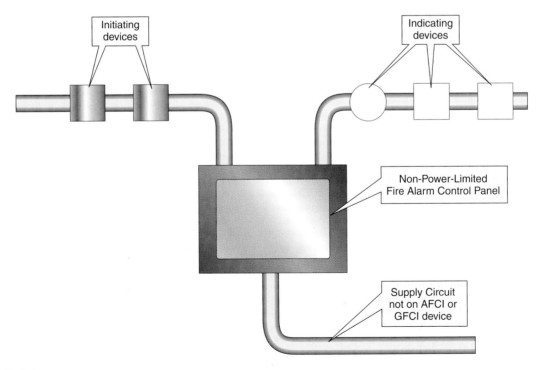

Initiating devices

Indicating devices

Non-Power-Limited Fire Alarm Control Panel

Supply Circuit not on AFCI or GFCI device

FIGURE 7.14

An interesting explanation of vote was made by one of the members of CMP-3, which has jurisdiction for Article 760: "The substantiation in both the proposal and the comment appears to be addressing concerns with AFCI protection for branch circuits supplying single- and multiple-station smoke detectors. These devices are self-contained assemblies that incorporate the detector, the control equipment, and the alarm-sounding device in one unit operated from a power supply either in the unit or obtained at the point of installation. Article 760 does not cover either single- or multiple-station detectors but rather addresses fire alarm systems employing a fire alarm panel.

Section 760.21 applies to the branch circuit supplying a fire alarm system and not to individual single- or multiple-station smoke detectors. Branch circuits supplying single- or multiple-station smoke detectors in a bedroom must comply with the requirements in 210.12."

The Technical Correlating Committee of the Signaling Systems for the Protection of Life and Property submitted a Comment supporting the Panel action on the proposal. However, the Technical Correlating Committee for the NEC did not take action on or restrict in any way this proposal. Perhaps we have not heard the end of this story.

760.41

Power Sources for Power-Limited Fire Alarm Circuits

NEC page 587

CHANGE TYPE: Revision (Proposal 3-256)

CHANGE SUMMARY: In this section on power-limited fire alarm circuits, the last sentence has been changed to require that "these circuits shall not be supplied through ground-fault circuit interrupters or arc-fault circuit-interrupters."

2005 CODE: "**760.41 Power Sources for PLFA Circuits.** The power source for a power-limited fire alarm circuit shall be as specified in 760.41(A), (B), or (C). These circuits shall not be supplied through ground-fault circuit interrupters <u>or arc-fault circuit interrupters</u>.

FPN NO. 1: Tables 12(A) and 12(B) in Chapter 9 provide the listing requirements for power-limited fire alarm circuit sources.

FPN NO. 2: See 210.8(A)(5), Exception No. 3, for receptacles in dwelling-unit unfinished basements that supply power for fire alarm systems."

CHANGE SIGNIFICANCE: However, the issue in this section differs from the revised rule for non-power-limited fire alarm circuits as this section applies to the power supply for *power-limited* fire alarm circuits. In reality, all power-limited fire alarm circuits originate at a fire alarm control panel. The prohibition for supply from a GFCI device or now from an AFCI device applies to the 120-volt circuit to the fire alarm control panel.

As revised for the 2005 NEC, 210.12 continues to require all outlets in the dwelling unit bedroom to be provided with AFCI protection.

FIGURE 7.15

This would include 120-volt outlets for fire alarm systems if the control panel is located in a dwelling unit bedroom.

However, the organization of the Code in 90.3 also applies to this issue. The rules in Chapter 2, where 210.12 is located, apply generally unless amended or modified by rules in Chapters 5, 6, or 7. Since Article 760 is in Chapter 7 of the Code, the rules in Article 760 can amend the rules in Article 2. As a result, the rule in 760.41 can be interpreted to act as an exception to the rule in 210.12 and prohibit the branch circuit for control panels supplying power-limited fire alarm circuits from being supplied by an AFCI-protected branch circuit.

The Technical Correlating Committee of the Signaling Systems for the Protection of Life and Property submitted a Comment supporting the Panel action on the proposal, as did the Automatic Fire Alarm Association.

760.61(A), 760.61(B), and 760.61(C)

Applications of Listed Power-Limited Fire Alarm Cables

NEC page 589

CHANGE TYPE: Revision (Proposals 3-273, 3-275, and 3-277)

CHANGE SUMMARY: These proposals add circuit integrity (CI) cable as a permissible wiring method to provide two-hour rated systems for fire alarm systems.

2005 CODE: "**760.61 Applications of Listed PLFA Cables.** PLFA cables shall comply with the requirements described in either 760.61(A), (B), or (C) or where cable substitutions are made as shown in 760.61(D).

(A) Plenum. Cables installed in ducts, plenums, and other spaces used for environmental air shall be Type FPLP. ~~Abandoned cables shall not be permitted to remain.~~ Types FPLP, FPLR, and FPL cables installed in compliance with 300.22 shall be permitted. <u>Type FPLP-CI cable shall be permitted to be installed to provide a two-hour circuit integrity rated cable.</u>

(B) Riser. Cables installed in risers shall be as described in either (1), (2), or (3):

(1) Cables installed in vertical runs and penetrating more than one floor, or cables installed in vertical runs in a shaft, shall be Type FPLR. Floor penetrations requiring Type FPLR shall contain only cables suitable for riser or plenum use. ~~Abandoned cables shall not be permitted to remain.~~

(2) Other cables shall be installed in metal raceways or located in a fire-proof shaft having firestops at each floor.

(3) Type FPL cable shall be permitted in one- and two-family dwellings.

FPN: See 300.21 for firestop requirements for floor penetrations.

FIGURE 7.16

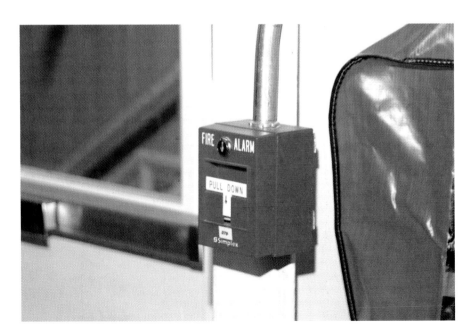

(C) Other Wiring Within Buildings. Cables installed in building locations other than those covered in 760.61(A) or (B) shall be as described in either (1), (2), (3), or (4). Type FPLP-CI cable shall be permitted to be installed as described in either (1), (2), (3), or (4) to provide a two-hour circuit integrity-rated cable.

(1) Type FPL shall be permitted.

(2) Cables shall be permitted to be installed in raceways.

(3) Cables specified in Chapter 3 and meeting the requirements of 760.82(A) and 760.82(B) shall be permitted to be installed in non-concealed spaces where the exposed length of cable does not exceed 3 m (10 ft).

(4) A portable fire alarm system provided to protect a stage or set when not in use shall be permitted to use wiring methods in accordance with 530.12."

CHANGE SIGNIFICANCE: The NEC Technical Correlating Committee, in its action on Comment 16-98 for the 2002 NEC, overturned CMP-16's acceptance of proposals to establish listing requirements for limited combustible cable, "because the Panel's action contains no requirements or specifications for the use of limited combustible cable versus the general cables already specified.... The Technical Correlating Committee notes that it is inappropriate to attempt to include references to all products that do not have a need for specific application rules or products that are permitted but not required by the NEC."

Type FPLP-CI cable was included in the 2002 NEC with appropriate listing and marking requirements, but no use requirements were provided. This proposed text responds to the NEC TCC ruling. NFPA 72-2002, the National Fire Alarm Code, has a two-hour circuit integrity requirement for certain circuits. One of the permitted options is a two-hour fire-rated cable. Therefore, circuit integrity cable with a "CI" suffix will remain in the NEC.

The Technical Correlating Committee for the Signaling Systems for the Protection of Life and Property submitted a Comment supporting the Panel's action on all three of these proposals.

CHAPTER 8

Communication Systems, Articles 800–830; Annexes

800.2

Communications Circuit Integrity (CI) Cable

NEC page 597

CHANGE TYPE: New (Proposal 16-74)

CHANGE SUMMARY: A new definition of *communications circuit integrity (CI) cable* has been added to 800.2 since the term is now used in Article 800.

2005 CODE: "**Communications Circuit Integrity (CI) Cable.** Cable used in communications systems to ensure continued operation of critical circuits during a specified time under fire conditions."

CHANGE SIGNIFICANCE: A heightened interest in the ability to maintain communications throughout the entire time of an emergency has prompted numerous agencies to require that steps be taken to assure the circuit integrity and survivability of certain critical communications circuits during a fire in a building. Additionally, designers and installers of electrical systems, including communications systems, continually seek performance guidance from the NEC in order to assure the safety of communications system installations. It is vitally important that the NEC offer a Code-complying method for meeting these requirements.

Circuit integrity was introduced in Article 760 in the 1999 Code, and given a common-sense definition that referred to a cable's capability "to ensure continued operation of critical circuits during a specified time under fire conditions." The corresponding FPN references UL 2196 as the required fire test.

FIGURE 8.1

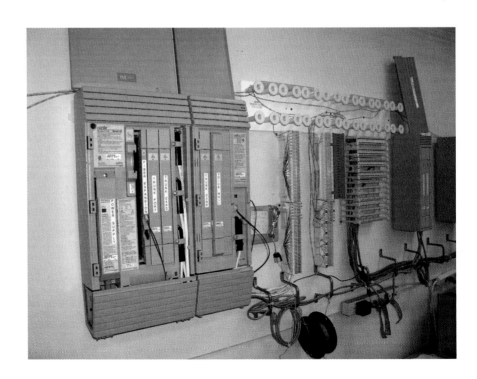

CHANGE TYPE: New (Proposal 16-96)

CHANGE SUMMARY: A new Fine Print Note has been added to point out that limiting the length of the primary protector grounding conductor at apartment and commercial buildings will help reduce voltages that may develop during lightning events.

2005 CODE: "**(4) Length.** The primary protector grounding conductor shall be as short as practicable. In one- and two-family dwellings, the primary protector grounding conductor shall be as short as practicable, not to exceed 6.0 m (20 ft) in length.

FPN: <u>Similar grounding conductor length limitations applied at apartment buildings and commercial buildings will help to reduce voltages that may be developed between the building's power and communications systems during lightning events.</u>"

CHANGE SIGNIFICANCE: When the above 20-foot limitation was instituted in the 2002 NEC, apartment and commercial buildings were specifically not addressed, as it was felt that the predominant issue regarding the length of the primary protector grounding conductor affected one- and two-family dwellings. In retrospect, some guidance is now provided for apartment and commercial buildings, without being overly restrictive because of intersystem bonding situations that may exist at these facilities.

The new FPN provides guidance for the treatment of the cable and primary protector grounding conductor length at apartment and commercial buildings, consistent with the 20-foot rule for one- and two-family dwellings, but does not place untenable restrictions on the actual length.

This was a companion proposal, intended to correlate with similar proposals for 820.100(A)(4) for CATV systems and for 830.100(A)(4) for network-powered broadband communications systems.

800.100 (A)(4) FPN

Primary Protector Grounding Conductor
NEC page 600

FIGURE 8.2

800.113

Installation and Marking of Communications Wires and Cables

NEC page 601

CHANGE TYPE: Revision (Proposal 16-101)

CHANGE SUMMARY: The listing requirements for communications cable have been reworded. Also, the previous Exception No. 2 has been deleted due to the other changes that have been made. Finally, a new Fine Print Note adds information on the length of unlisted plant cable inside a building.

2005 CODE: "**800.113 Installation ~~Listing,~~ and Marking, ~~and Installation~~ of Communications Wires and Cables.** Listed communications wires and cables <u>and listed multipurpose cables shall be</u> installed as wiring within buildings ~~shall be listed as being suitable for the purpose and installed in accordance with 800.52.~~ Communications cables and undercarpet communications wires shall be marked in accordance with Table 800.113. The cable voltage rating shall not be marked on the cable or on the undercarpet communications wire.

FPN: Voltage markings on cables may be misinterpreted to suggest that the cables may be suitable for Class 1, electric light, and power applications.

Exception No. 1: Voltage markings shall be permitted where the cable has multiple listings and voltage marking is required for one or more of the listings.

~~*Listing and marking shall not be required where the cable enters the building from the outside and is continuously enclosed in a rigid metal conduit system or an intermediate metal conduit system and such conduit systems are grounded to an electrode in accordance with 800.40(B).*~~

FIGURE 8.3

Exception No. 2: ~~Exception No. 3:~~ Listing and marking shall not be required where the length of the cable within the building, measured from its point of entrance, does not exceed 15 m (50 ft) and the cable enters the building from the outside and is terminated in an enclosure or on a listed primary protector.

FPN NO. 1 TO EXCEPTION NO. 2:   Splice cases or terminal boxes, both metallic and plastic types, are typically used as enclosures for splicing or terminating telephone cables.

FPN NO. 2 TO EXCEPTION NO. 2:   This exception limits the length of unlisted outside plant cable to 15 m (50 ft), while 800.30(B) requires that the primary protector be located as close as practicable to the point at which the cable enters the building. Therefore, in installations requiring a primary protector, the outside plant cable may not be permitted to extend 15 m (50 ft) into the building if it is practicable to place the primary protector closer than 15 m (50 ft) to the entrance point."

CHANGE SIGNIFICANCE: Previously, listing requirements were not presented uniformly in similar articles. This proposal makes the NEC more user friendly in that all listing requirements will be included in a new section 725.82, 760.81, 770.179, 800.170, 820.179, and 830.179. This revision is similar to how the listing requirements for Class 2, Class 3, and fire alarm circuits were moved from Articles 725 and 760 to Chapter 9. Also, CMP-16 renumbered the articles so that like installation requirements have similar section numbers whenever possible.

This proposal revises Section 800.50 such that it covers installation requirements only (listing requirements are covered elsewhere). Furthermore, the previous Exception No. 2 was deleted because it became redundant when the definition of *point of entrance* was added to the 1999 NEC.

These changes require listed cables to be installed inside buildings. An existing exception permits up to 50 feet of unlisted, outside plant communications cable to be installed in limited spaces.

Annex D, Example D3(a)

D3(a) Industrial Feeders in a Common Raceway
NEC page 719

CHANGE TYPE: New (Proposal 2-367)

CHANGE SUMMARY: A new load calculation example of industrial feeders in a common raceway has been added. It attempts to set out integration of the rules on continuous and noncontinuous loading, terminations, ampacity conditions, and environmental influences. It also deals with the conditions of the heating effects of the interior of raceways and terminations.

2005 CODE: D3(a) Industrial Feeders in a Common Raceway
An industrial multi-building facility has its service at the rear of its main building, and then provides 480Y/277-volt feeders to additional buildings behind the main building in order to segregate certain processes. The facility supplies its remote buildings through a partially enclosed access corridor that extends from the main switchboard rearward along a path that provides convenient access to services within 15 m (50 ft) of each additional building supplied. Two building feeders share a common raceway for approximately 45 m (150 ft) and run in the access corridor along with process steam and control and communications cabling. The steam raises the ambient temperature around the power raceway to as much as 35°C. At a tee fitting, the individual building feeders then run to each of the two buildings involved. The feeder neutrals are not connected to the equipment grounding conductors in the remote buildings. All distribution equipment terminations are listed as being suitable for 75°C connections.

Each of the two buildings has the following loads:

Lighting, 11,600 VA, comprised of electric-discharge luminaires connected at 277 V

Receptacles, 22 125-volt, 20 ampere receptacles on general-purpose branch circuits, supplied by separately derived systems in each of the buildings

1-Air compressor, 460 volt, three phase, 7.5 hp

1-Grinder, 460 volt, three phase, 1.5 hp

3-Welders, AC transformer type (nameplate: 23 amperes, 480 volts, 60 percent duty cycle)

3-Industrial Process Dryers, 480 volt, three phase, 15 kW each (assume continuous use throughout certain shifts)

Determine the overcurrent protection and conductor size for the feeders in the common raceway, assuming the use of XHHW-2 insulation (90°C):

Calculated Load {Note: For reasonable precision, volt-ampere calculations are carried to three significant figures only; where loads are converted to amperes, the results are rounded to the nearest ampere [see 220.5(B)].}

Noncontinuous Loads
Receptacle Load (see 220.44)
22 receptacles at 180 VA   3,960 VA
Welder Load [see 630.11(A), Table 630.11(A)]
Each welder: 480 V × 23A × 0.78 = 8,610 VA
All 3 welders: [see 630.11(B)]
   (demand factors 100%, 100%, 85% respectively)
8,610 VA + 8,610 VA + 7,320 VA =   24,500 VA

Subtotal, Noncontinuous Loads   **28,500 VA**

Motor Loads (see 430.24, Table 430.250)
Air compressor: 11 A × 480 V × √ 3 =   9,150 VA
Grinder: 3 A × 480 V × √ 3 =   2,490 VA
Largest motor, additional 25%:   2,290 VA

Subtotal, Motor Loads   **13,900 VA**

By using 430.24, the motor loads and the noncontinuous loads can be combined for the remaining calculation.

Subtotal for load calculations, Noncontinuous Loads   **42,400 VA**

Continuous Loads
General Lighting   11,600 VA
3 Industrial Process Dryers   15 kW each   45,000 VA

Subtotal, Continuous Loads: **56,600 VA**

Overcurrent Protection (*see* 215.3.)
The overcurrent protective device must accommodate 125% of the continuous load, plus the noncontinuous load:

Continuous load	56,600 VA
Noncontinuous load	42,400 VA
Subtotal, actual load [actual load in amperes]:	**99,000 VA**

99,000 VA ÷ (480 V × $\sqrt{3}$) = 119 A

(25% of 56,600 VA) (*see* 215.3.)	14,200 VA
Total VA	**113,200 VA**

Conversion to amperes using three significant figures:
   113,200 VA / (480 V × $\sqrt{3}$) = 136 A
Minimum size overcurrent protective device: 136 A
Minimum standard size overcurrent protective device (*see* 240.6):
   150 amperes

Where the overcurrent protective device and its assembly are listed for operation at 100 percent of its rating, a 125 ampere overcurrent protective device would be permitted. However, overcurrent protective device assemblies listed for 100 percent of their rating are typically not available at the 125-ampere rating. (*See 215.3 Exception.*)

Ungrounded Feeder Conductors
The conductors must independently meet requirements for (1) terminations, and (2) conditions of use throughout the raceway run.

Minimum size conductor at the overcurrent device termination [*see 110.14(C) and 215.2(A)(1), using 75°C ampacity column in Table 310.16*]: 1/0 AWG.

Minimum size conductors in the raceway based on actual load [*see Article 100, Ampacity, and 310.15(B)(2)(a) and correction factors to Table 310.16*]:

$$99,000 \text{ VA} / 0.7 / 0.96 = 147,000 \text{ VA}$$

(70% = 310.15(B)(2)(a)) & (0.96 = Correction factors to Table 310.16)

Conversion to amperes:

$$147,000 \text{ VA} / (480 \text{ V} \times \sqrt{3}) = 177 \text{ A}$$

Note that the neutral conductors are counted as current-carrying conductors [*see 310.15(B)(4)(c)*] in this example because the discharge lighting has substantial nonlinear content. This requires a 2/0 AWG conductor based on the 90°C column of Table 310.16. Therefore, the worst case is given by the raceway conditions, and 2/0 AWG conductors must be used. If the utility corridor was at normal temperatures [30°C (86° F)], and if the lighting at each building were supplied from the local separately derived system (thus requiring no neutrals in the supply feeders) the raceway result (99,000 VA / 0.8 = 124,000 VA; 124,000 VA / (480 V × $\sqrt{3}$) = 149 A, or a 1 AWG conductor @ 90°C) could not be used because the termination result (1/0 AWG based on the 75°C column of Table 310.16) would become the worst case, requiring the larger conductor.

In every case, the overcurrent protective device must provide overcurrent protection for the feeder conductors in accordance with their ampacity as provided by this *Code* (*see* 240.4). A 90°C 2/0 AWG conductor has a Table 310.16 ampacity of 195 amperes. Adjusting for the conditions of use (35°C ambient temperature, 8 current-carrying conductors in the common raceway),

$$195 \text{ amperes} \times 0.96 \times 0.7 = 131 \text{ A}$$

The 150-ampere circuit breaker protects the 2/0 AWG feeder conductors, because 240.4(B) permits the use of the next higher standard size overcurrent protective device. Note that the feeder layout precludes the application of 310.15(A)(2) Exception.

Feeder Neutral Conductor (*see 220.61*):
Because 210.11(B) does not apply to these buildings, the load cannot be assumed to be evenly distributed across phases. Therefore the maximum imbalance must be assumed to be the full lighting load in this case, or 11,600 VA. (11,600 VA / 277 V = 41 amperes.) The ability of the neutral to return fault current [*see 250.32(B)(2)(2)*] is not a factor in this calculation.

Although the neutral runs between the main switchboard and the building panelboard, likely terminating on a busbar at both locations, the busbar connections are part of listed devices and are not "separately installed pressure devices." Therefore 110.14(C)(2) does not apply, and the normal termination temperature limits apply. In addition, the listing requirement to gain exemption from the additional sizing allowance under continuous loading (*see 215.3 Exception*) covers not just the overcurrent protective device, but its entire assembly as well. Therefore, since the lighting load is continuous, the minimum conductor size is based on 1.25 × (11,600 VA / 277 V) = 52 amperes, to be evaluated under the 75°C column of Table 310.16. The minimum size of the neutral is 6 AWG. This size is also the minimum size required by 215.2(A)(1), because the minimum size equipment grounding conductor for a 150 ampere circuit, as covered in Table 250.122, is 6 AWG.

CHANGE SIGNIFICANCE: This proposal and several others related to Article 220 were developed under the guidance of the NFPA task group on usability. Several sub-task groups were formed to evaluate the need for a reorganization of Article 220 and to determine where editorial revisions would enhance the usability of the article.

This proposal provides a new example that will greatly enhance code usability by providing actual integration of rules covering continuous/noncontinuous loading, terminations, ampacity conditions, and environmental influences. The present Example D3 addresses this slightly, but does not cover three-phase industrial systems and the complexities inherent when considering ambient temperature and mutual conductor heating in the context of other rules.

The key to the new approach is to clearly set off the rules that follow from the effects of heating in the interior of a raceway or cable assembly, and which are solely intended to protect the integrity of the insulation, from those that apply at terminations. Termination rules have little to do with protecting conductor insulation and everything to do with assuring proper function of devices, many of which will not function properly absent a properly sized heat sink bolted to them.

In short, every conductor has both a middle and an end. Different thermodynamic considerations apply to each, even if certain derating factors may coincidentally be equal. For example, 80 percent may apply to continuous load profiles to assure appropriate termination performance, and the same percentage may apply to six current-carrying conductors in a raceway. This is purely coincidental, as is obvious if one jumps to nine such conductors. Nevertheless, there is widespread confusion on these points in the field.

Every Code Making Panel is faced with the dilemma of writing Code that is simple to read and apply, but that overdesign electrical installations and waste resources in the process, and writing technically correct Code that is more difficult to read and apply uniformly. One way to bridge this gap is through the judicious use of examples. This example fulfills that need.

CHANGE TYPE: Revision and New (Proposal 7-212a)

CHANGE SUMMARY: The new Table E.2 added to Annex E gives the maximum number of stories permitted for building construction Type III, IV, and V, including where various construction methods are used in the different construction-type buildings. Additional text is added to describe the various construction types.

2005 CODE: Add the following to Annex E, following the table in the existing Annex.

The five different types of construction can be summarized briefly as follows (see also Table E.2)

Type I is a Fire-Resistive construction type. All structural elements and most interior elements are required to be noncombustible. Interior, nonbearing partitions are permitted to be 1 or 2 hour rated. For nearly all occupancy types, Type 1 construction can be of unlimited height.

Type II construction has 3 categories: Fire-Resistive, One-Hour Rated, and Non-Rated. The number of stories permitted for multifamily dwellings varies from 2 for Non-Rated, and 4 for One-Hour Rated to 12 for Fire-Resistive construction.

Type III construction has two categories, One-Hour Rated and Non-Rated. Both categories require the structural framework and exterior walls to be of noncombustible material. One-Hour Rated construction requires all interior partitions to be one-hour rated. Non-Rated construction allows nonbearing interior partitions to be of non-rated construction. The maximum permitted number of stories for multifamily dwellings and other structures is 2 for Non-Rated and 4 for One-Hour Rated.

Type IV is a single construction category which provides for heavy timber construction. Both the structural framework and the exterior walls are required to be noncombustible, except that wood members of certain minimum sizes are allowed. This construction type is seldom used for multifamily dwellings but, if used, would be permitted to be 4 stories high.

Type V construction has two categories, One-Hour Rated and Non-Rated. One-Hour Rated construction requires a minimum of one-hour rated construction throughout the building. Non-rated construction allows non-rated interior partitions with certain restrictions. The maximum permitted number of stories for multifamily dwellings and other structures is 2 for Non-Rated and 3 for One-Hour Rated.

CHANGE SIGNIFICANCE: A new table is being added that gives the maximum number of stories permitted for construction Types III, IV, and V. In addition, text is being added to explain the general characteristics of the five construction types.

Annex E
Types of Construction
NEC page 726

INDEX

Ground-fault circuit interrupters (GFCI)
  agricultural buildings, receptacles, 308
  aircraft hangers, 286
  boat hoists, outlets for, 50–51
  carnivals, circuses, fairs, and similar events, 306–307
  dwelling units, 44, 50–51
  elevator car lighting, 326
  HVAC equipment, receptacles installed for, 49
  kitchens, commercial and institutional, 45–46
  luminaires protected by, 224–225
  natural and artificially made bodies of water, electrical installations, 348
  outdoor public spaces, receptacles in, 47–48
  storable swimming pools, 342
  switches, grounding of general-use snap, 209–210
  vending machines, cord-and-plug connected, 237–238
Ground-fault current path, effective, definition, 102
Ground-fault detectors, 102
Ground-fault protection
  fire pumps, 364–365
  health care facilities, 296–297
  supply side of service disconnect, connected to, 86
Grounding and bonding, 101. *See also* Bonding jumpers; Equipment grounding; Equipment grounding conductors; Grounding electrode conductors; Grounding electrode system; Grounding electrodes
  alternating-current system of 50 volts to 1000 volts, 104–105
  autotransformers, 261
  buildings or structures supplied by feeders or branch circuits, 113–114
  carnivals, circuses, fairs, and similar events, 304–305
  connection of equipment, 103
  cranes and hoists, 325
  effective ground-fault current path, definition, 102
  equipotential bonding
    agricultural buildings, 311–312
    naturally and artificially made bodies of water, electrical installations, 348–349
    swimming pools and similar installations, 338–341
  hazardous (classified) locations, bonding in, 127
  hydromassage bathtubs, 344
  luminaires, exposed parts, 224–225
  meter enclosures on load side of service disconnect, 134
  multipoint grounding, 138–140
  natural and artificially made bodies of water, electrical installations, 348

1 kV and over systems and circuits, solidly grounded systems, 138–140
  patient care areas, receptacles and fixed electric equipment in, 294–295
  photovoltaic systems, ac and dc requirements, 356–357
  separately derived alternating-current systems, 108–112
  single point grounding, 138–140
  switches, general-use snap, 209–210
Grounding conductors
  connection of, 103
  primary protector, communications circuits, 395
Grounding electrode conductors
  common, 121–122
  connection to grounding electrodes, accessibility, 125–126
  definition, 15
  enclosures for, 123–124
  installation, continuous, 119–120
  photovoltaic systems, ac and dc requirements, 356–357
  separately derived system, 109–110, 112
  taps, 121–122
Grounding electrode system
  bonding together of grounding electrodes, 115–116
  buildings or structures supplied by feeders or branch circuits, 113–114
  carnivals, circuses, fairs, and similar events, multiple sources of supply, 304–305
  concrete-encased electrodes, 115–116
Grounding electrodes
  definition, 14
  grounding electrode conductor and bonding jumper connection to, accessibility, 125–126
  metal frame of building or structure as, 117–118
  photovoltaic systems, ac and dc requirements, 356–357
  separately derived alternating-current systems, 111–112
Guest rooms or guest suites, 38
  branch circuits, 54
  cooking facilities in, 60–61
  definitions, 16–17
  lighting outlets, 63–64
  receptacle outlets, 60–61
Gutters, auxiliary, conductors in, 165–166

Handhole enclosures, 170–171, 176–177
  conductors in, boxes or conduits in, 153–154
  definition, 18
Hangers, aircraft. *See* Aircraft hangars
Hazardous (classified) locations
  bonding in, 127